Sensory Evaluation in Quality Control

Sensory Evaluation in Quality Control

Alejandra M. Muñoz
Sensory Spectrum, Inc.
Gail Vance Civille
Sensory Spectrum, Inc.
B. Thomas Carr
The NutraSweet Co. R&D

VNR SPRINGER SCIENCE+BUSINESS MEDIA, LLC

Copyright © 1992 by Springer Science+Business Media New York
Originally published by Van Nostrand Reinhold in 1992
Softcover reprint of the hardcover 1st edition 1992

Library of Congress Catalog Card Number 91-26647
ISBN 978-1-4899-2655-5

16 15 14 13 12 11 10 9 8 7 6 5 4 3 2~1

Library of Congress Cataloging-in-Publication Data

Muñoz, Alejandra M., 1957–
    Sensory evaluation in quality control / Alejandra M. Muñoz. Gail
Vance Civille, B. Thomas Carr.
        p.   cm.
    "An AVI book."
    Includes index.
    ISBN 978-1-4899-2655-5   ISBN 978-1-4899-2653-1 (eBook)
    DOI 10.1007/978-1-4899-2653-1

    1. Quality control.   2. Sensory evaluation   I. Civille, Gail
Vance.   II. Carr, B. Thomas.   III. Title.
TS156.M86   1991
658.5'62—dc20   91-26647
                                                                        CIP

To Michael, Frank, and Cathy

Contents

Preface

This book addresses an important, but so far neglected, topic: the application of sensory evaluation to quality control. Although several articles have been published that have discussed concepts of quality control/sensory evaluation (QC/sensory) programs, *Sensory Evaluation in Quality Control* is the first publication that addresses this topic in a comprehensive and practical way. This book is comprehensive, in that it presents the sensory and statistical information that is needed to design and implement several types of QC/sensory programs at the plant level. The book is practical, in that it provides a step-by-step description of the complete process to implement such programs, and it illustrates this process through real examples encountered by various consumer products companies (e.g., foods, personal care products, paper products). With this practical information, sensory and quality professionals can design and implement sound QC/sensory programs at the plant level.

This book was developed to provide the sensory and quality professional with an overview and guide to apply, in a production facility, the unique techniques that are used to measure sensory responses. Therefore, the book is intended for QC and/or R&D personnel (e.g., sensory managers and analysts, and quality professionals) in charge of implementing an in-plant program, as well as for the plant management and plant technical personnel (sensory coordinator and quality professionals) who are ultimately responsible for the routine operation of the established program. Technical personnel can use this book as a guide and benefit mainly from the in-depth description of the sensory and statistical information. QC, R&D, and plant management can use it to assess the benefits of a sensory program at the plant level, the various alternative techniques that exist, the advantages and limitations of each, and the resources and costs needed to implement any of the QC/sensory programs discussed.

This book is organized into four main parts. Chapter 1 is an overview of quality control programs, of the test methods in a QC operation (instrumental and sensory), and of the common sensory techniques that are published in the sensory literature and used in industry, with a discussion of the advantages and disadvantages of each. Chapter 2 is a discussion of the preliminary phases in the program implementation, such as the planning phase and the assessment of resources and needs. Chapters 3 to 6 present the in-depth description of four different QC/sensory programs: comprehensive descriptive, quality ratings, "in/out," and difference from control methods. Finally, the appendices contain statistical background information on the procedures, for summarizing and analyzing QC/sensory data.

Each program (chapters 3 to 6) is covered in four basic sections: 1) an abstract, 2) a description of the program in operation, 3) the steps and procedures for implementing and operating the program, and 4) other versions of the program. The abstract contains a brief summary of the method and describes the main characteristics of the method, as well as its uses, advantages, and disadvantages. The section on operation of the program describes the way the program functions. It shows the type of data collected, the type of sensory specification used, and the comparison of the collected data and the specifications for product decisions. This section allows the reader to become familiar with the methodology and principles of the program. The section on implementation of the program provides detailed information on each step recommended for developing and implementing such a program. This section is intended to provide the reader with information needed to implement that QC/sensory program. Finally, the section on other versions of the program presents modified programs of the main method. In many situations, due to limitations on resources or philosophical differences, professionals might opt to consider implementing modified or shorter versions instead of the complete program discussed in each chapter.

Each method has unique characteristics and approaches that are discussed in their corresponding chapter. There are, however, many common components and tests among all methods. These common elements are discussed in detail only in the first method presented—the comprehensive descriptive approach (Chapter 3). The other chapters (4–6) make frequent reference to the common program aspects discussed in Chapter 3.

Our special appreciation is extended to Barbara Pirmann and Andrea Senatore, for preparing and proofing this manuscript. We wish to also thank Clare Dus, for her assistance in the literature review, and Linda Brands, for her help in preparing several tables and figures of the book.

Alejandra M. Muñoz
Gail Vance Civille
B. Thomas Carr

1

Introduction

THE CONCEPT OF QUALITY

The Importance of Quality in American Industry

In today's very competitive and dynamic marketplace, thousands of new products and services are being offered to consumers on a regular basis. Consumers are faced with a variety of alternatives from which to choose and are forced to assess the product's factors (e.g., quality, cost), when making their product selection. Some consumers consider quality the most important factor while; for others it is cost or other parameters, such as availability, convenience, and so forth.

In the 1900s, there was little information on the link between quality and consumers' buying decisions. However, in the 1960s when the Japanese first began to sell their high-quality products in several U.S. markets with evident success, it was confirmed that consumers *do care for quality and buy quality* (Plsek 1987; Pedraja 1988). Other information supports the importance of quality in a product's success. For example, a variety of business surveys indicate that gains in market share and growth have been greater in companies that strive for excellent quality, than for those that produce low-quality products (Gallup Organization 1985).

Studies have shown that American consumers discriminate among quality products and prefer the highest and most reliable product option, such as the Japanese product counterparts (Schrock and Lefevre 1988; Penzias 1989). Garvin (1987) reports the results of a 1981 survey that found that nearly 50 percent of U.S. consumers believed that the quality of U.S. products had dropped in the previous five years. In addition, 25 percent of the consumers were "not at all" confident that U.S. industry can be depended on to deliver reliable products.

Schrock and Lefevre (1988) pose the question, "When will the United States learn?" and indicate that the United States has an outstanding ability to produce

1

complex and sophisticated products, but that this is insufficient. Efforts are needed for the country to regain its world leadership in quality. W. Edward Deming, who taught Japan about quality, states that, "We in America will have to be more protectionist or more competitive. The choice is very simple: if we are to become more competitive, then we have to begin with our quality" (Halberstam 1984).

In light of this scenario, American companies have committed to establishing programs for the development, maintenance and/or improvement of quality products. Companies know that the total cost of quality programs is outweighed by the benefits they produce. These benefits include savings, in terms of materials, effort, and time, as well as the enhanced business resulting from higher consumer acceptance and greater competitiveness.

Currently, many companies consider the pursuance and maintenance of quality an essential part of their operation. Experts consider that, without quality programs, companies are doomed to fail. Cullen and Hollingum (1987) believe that, in the future, there will be two types of companies: those that have implemented total quality and those that will be out of business. At the same time, these experts consider the Japanese quality perspective to be the model to follow for success. (Hayes 1981; Halberstam 1984; Cullen and Hollingum 1987; Schrock and Lefevre 1988; Penzias 1989).

The problem and the challenge are very complex. Within a company, the pursuit of quality must be the job and challenge of each employee, from top management to line workers. Educational and operational quality programs and the concepts of quality should be an essential component of the corporate culture. Many quality control books cover, in detail, endeavors that many American consumer products companies have undertaken: programs developed to establish and maintain their products' quality. This book adds a new tool to quality control programs that is critical for the consumer products industry: evaluating and controlling products, based on their sensory properties.

Definitions of Quality

Sinha and Willborn (1985) present a collection of definitions of quality (Table 1-1). Intrinsic in all these definitions is the concept that quality encompasses the characteristics of a product or service that are designed to meet certain needs under specified conditions. Although these are accurate definitions that are widely used and sometimes sufficient for discussing general concepts, there are additional aspects that need to be included, to better understand the total quality view. These concepts are:

- The consumer input in the definition of quality;
- The multidimensional nature of quality; and
- Quality consistency.

Table 1-1.   General definition of quality

Definition	Source
"The totality of features and characteristics of a product or service that bear on its ability to satisfy given needs."	1, 4, 5
"Fitness for use."	2
"Conformance to requirements."	3
"The degree to which product characteristics conform to the requirements placed upon that product, including reliability, maintainability, and safety."	6
"The degree to which a product or service is fit for the specified use."	7

1. ANSI/ASQC Standard, "Quality System Terminology," A3-1978, prepared jointly by the American National Standards Institute (ANSI) and American Society for Quality Control (ASQC).
2. Juran, J. M., editor-in-chief, *Quality Control Handbook*, 3rd. ed., McGraw Hill Book Company, New York, 1974.
3. Crosby, P.B., *Quality Is Free*. McGraw Hill Book Company, New York, 1979.
4. DIN-53350. of Deutsches Institute fuer Normung Teil 11, Beuth-Verlag, Berlin.
5. EOQC, "Glossary of Terms Used in Quality Control," 5th ed., 1981, published by the European Organization for Quality Control.
6. QS-Norm Draft of Swiss Standard Association, 1981.
7. Seghezzi, H.D., "What is Quality Conformance With Requirements or Fitness for the Intended Use," *EOQC Journal* 4, 1981, p.3.
Source: Sinha and Willborn 1985.

The Consumer Input in the Definition of Quality

McNutt (1988) emphasizes the importance of defining quality in terms of consumer perceptions. He points out that the definition and assessment of a product's quality might be different for industry's management and the consumer. Garvin (1987) cites a survey that shows that 65 percent of business executives thought consumers could readily name a good quality brand in a big-ticket product category. When asked, only 16 percent of consumers could name such a brand for small appliances and only 23 percent could give a name for large appliances.

If the generic definitions of quality imply the satisfaction of needs, it should be made clear that it is the person *whose needs are being satisfied* who determines quality—that is, the consumer. Thus:

- The consumer's input is required to determine the product's quality parameters.
- It is the consumer's perception that underlies his or her assessment of quality.
- The consumer's perception of quality cannot be predicted.
- A complete study is required, to fully understand consumer perceptions.

- Products and services consist of a variety of perceivable characteristics; therefore, it is important to determine which characteristics are important to consumers, in order to define quality.

Few quality control publications deal with the consumer dimension within an in-plant program. In an early publication, MacNiece (1958) explained the importance of studying the consumer requirements and integrating them into engineering specifications. Many industrial and commercial enterprises believe that they know what is best for consumers and what they *need*. However, companies need to understand the difference between what consumers "want" and what they "need" (MacNiece 1958). This book emphasizes the importance of consumer input in quality programs and includes specific procedures, to incorporate consumer responses in establishing quality programs.

The Multidimensional Nature of Quality
The unidimensional nature of the common definitions of quality (Table 1-1) is limiting (Plsek 1987). If quality is viewed as a multidimensional subject, differences and similarities among products and their different performances in the market can be better understood. All of a product's characteristics need to be considered when defining quality from the consumer perspective (McNutt 1988).

To better understand the quality of a product or service, Plsek (1987) expanded on Garvin's (1984) multidimensional perspective, by developing a nine-dimensional consideration of quality (Table 1-2). Following this philosophy, Table 1-3 shows two examples of the various quality dimensions for a product. This quality view implies that:

- Products or services are defined by a variety of characteristics.
- Because of the above, the quality of products is a multidimensional and complex problem.
- Products and services in the same category differ from each other in the way that they achieve excellence in different quality dimensions.

Given the multidimensional aspects of quality, the perceived quality of a new product (product's inherent quality) or service is determined by two main aspects:

1. The ability of a company to determine the critical quality dimensions of the product category at the initial stages of the product development; and
2. The ability of a company to achieve excellence in these critical quality dimensions.

The critical quality dimensions of a product are not stagnant. Their importance

Table 1-2.   The dimensions of quality

Dimension	Meaning
Performance	Primary product or service characteristics
Features	Added touches, "bells and whistles," secondary characteristics
Conformance	Match with specifications, documentation, or industry standards
Reliability	Consistency of performance over time
Durability	Useful life
Serviceability	Resolution of problems and complaints
Response	Characteristics of the human-to-human interface like timeliness, courtesy, professionalism, etc.
Aesthetics	Sensory characteristics like sound, feel, look, etc.
Reputation	Past performance and other intangibles

Source: Plsek 1987.

Table 1-3.   Examples of dimensions of quality

Dimension	Product Example: Stereo Amplifier	Service Example: Bank Checking Account
Performance	Signal-to-noise ratio, power	Time to process customer requests
Features	Remote control	Automatic bill paying
Conformance	Workmanship	Accuracy
Reliability	Mean-time-to-failure	Variability of time to process requests
Durability	Useful life (with repair)	Keeping pace with industry trends
Serviceability	Ease of repair	Resolution of errors
Response	Courtesy of dealer	Courtesy of teller
Aesthetics	Oak finished cabinet	Appearance of bank lobby
Reputation	*Consumer Reports* ranking	Advice of friends, years in business

Source: Plsek 1987.

to consumers can vary, as a result of very diverse phenomena, such as dynamic, social, and economic factors. Furthermore, companies are able to successfully alter the importance that consumers give to each of the product's quality dimensions, by means of aggressive advertisement.

The perceivable characteristics of a product, as measured by sensory methods, are multidimensional in nature. The methods presented in this book are akin to the multidimensional quality perspective of Garvin and Plsek. Specifically:

- The multidimensional nature of products is incorporated in the QC/sensory methods proposed.
- Methods for determining the effect of multidimensional variability on consumer acceptance are presented.

- Techniques for identifying the critical quality dimensions of a product, based on the consumer input, are discussed.
- The proposed method culminates in cost-effective quality programs that focus on only the critical quality dimensions of products.

Quality Consistency

The two quality dimensions, discussed earlier—the consumer input and the regard to the multidimensional product nature—play an important role in the first of the two stages of a product cycle, which is the development of the product and the establishment of its inherent quality. In this first product stage, the inherent quality characteristics are established and are meant to satisfy consumer needs and expectations. These inherent characteristics are, among others, the dimensions, aesthetics, performance, and cost of the product, and will determine the initial consumer acceptance or liking of a product. In the second product stage, the consistency aspect of quality plays an important role. The two product stages have to be considered in order to determine the total product quality. A quality product has inherent properties of high quality and will consistently deliver these quality properties to consumers, over time.

This book does not deal with issues related to the first product stage, which includes selecting the critical inherent quality parameters, establishing the inherent product quality, and performing the technical and business efforts during the product's development. This book's material deals with the quality issues that follow the development and production of a product of a given quality, which is the consistency aspect of quality. To achieve consistent high quality, a program for maintaining the inherent product's quality is required.

THE MAINTENANCE OF PRODUCT QUALITY (CONSISTENCY)

The Value of Product Consistency

Establishing the product's inherent quality is a very critical business and technical product phase since it is the basis for the future success of the product or service. Steven Jobs, of Apple Computer fame, (Gendron and Burlingham 1989) has the fundamental belief that time has to be sacrificed, in order to produce and launch a good product the first time. This is easier than having to go back and fix it. However, the success of a product (i.e., the consumer's continued acceptance) does not depend solely on this first product stage, but on the product's ability to fulfill a second need for consumer satisfaction: consistency. As stated by Lehr (1980), the definition of quality should encompass "the consistent conformance to customers' expectations." This implies "doing it right every time instead of just doing it right the first time."

The efforts and the responsibilities to maintain and control the product's quality shift within the company. While Product Development and Marketing/Market Research are in charge of establishing the initial product quality, Quality Control and Quality Assurance are responsible for establishing a system that assures the product's consistency.

The Function and Establishment of Quality Control Programs

There are a variety of definitions of quality control and quality assurance (see Table 1-4). Intrinsic in all these definitions is the operation of a system/program that compares items or services to standards and assures that items or services meet requirements, thus assuring consumer/customer satisfaction.

Marketing a high-quality product does not stop at the development phase, but extends into the manufacturing phase. The value of a product with high inherent quality characteristics has little importance if a company is unable to maintain and consistently deliver high quality to consumers. Therefore, it is essential that companies establish a structured function to measure the product's conformance to quality during its manufacture, and to control the product that is shipped to consumers. This function within a company is performed by a quality control group. In earlier days, the responsibility for quality was that of the workman and his foreman. As companies and the perception of the importance of quality grew, the responsibility of quality became more widely distributed. Quality became "everybody's job," but, consequently, was often nobody's job (Feigenbaum 1951). As a result, quality control functions were developed to coordinate this major activity within an organization.

In general, quality control benefits the target user and the company. The target user gets consistent quality, constant cost, and satisfaction from the products manufactured with an adequate quality control. The company achieves consumer satisfaction and loyalty, increased sales, high profits, and internal satisfaction among employees because of high-quality products produced, among other benefits.

The Steps in the Implementation of a QC Program

In general, there are five steps required to establish and manage a quality control program (Feigenbaum 1951; Caplen 1969):

1. Set standards/specifications—Determine the required quality costs and quality performance standards for the product.

Table 1-4.   Generic definitions of quality control/assurance

Definition	Source
Quality Control	
"An effective system for coordinating the quality maintenance and quality improvement efforts of the various groups in an organization so as to enable production at the most economical levels which allow for full consumer satisfaction."	1
"The core activity in any quality management program through which actual quality performance can be measured, compared against a qualified standard, and acted upon to prevent deficiencies."	2
Quality Assurance	
"All those planned or systematic actions necessary to provide adequate confidence that a product or service will satisfy given needs."	3
"A planned and systematic pattern of all means and actions designed to provide adequate confidence that items or services meet contractual and jurisdictional requirements and will perform satisfactorily in service. Quality Assurance includes Quality Control."	4

1. Feigenbaum, A.V. 1951. *Quality Control. Principles, practice and administration*. New York: McGraw-Hill Book Company.
2. Sinha, M.N. and W.O. Willborn. 1985. *The management of quality assurance*. New York: John Wiley and Sons.
3. ANSI/ASQC Standard, "Quality System Terminology," A3-1978. Prepared jointly by the American National Standards Institute (ANSI) and American Society for Quality Control (ASQC).
4. Canadian Standard Association Standard, Z299.1-1978. "Quality Assurance Program Requirements."

2. Prepare to manufacture, manufacture, and inspect—Decide on the materials and process. Inspect or control the operation during manufacture.
3. Appraise conformance—Assess the conformance of the manufactured product to the standards.
4. Act when necessary—Take corrective action when the standards are exceeded.
5. Plan for improvement—Develop a continuing effort to improve the cost and performance standards.

This book focuses on the role that sensory techniques can play in satisfying these five criteria.

The QC Steps that Require the Measurement of Product Characteristics

The five steps required to establish a QC program were listed above (Feigenbaum 1951; Caplen 1969). From these, the steps that involve measuring a product's properties are:

- Setting standards/specifications;
- Inspecting during or after manufacture;
- Appraising conformance; and
- Planning for improvements.

Setting standards/specifications—Establishing specifications requires a comprehensive and detailed knowledge of the properties of a product, namely:

- An in-depth knowledge of the attributes of the product manufactured;
- Knowledge of raw material components, their variability, and synergistic effects on the finished product;
- Knowledge of those product attributes (subset of above) that vary during production;
- Knowledge of the relationship between production variability and consumer acceptance; and
- Identification of the tolerable limits (levels/intensities) of product variability, in each attribute. These limits are the actual product specifications.

Each step requires measurement techniques to generate the necessary product or consumer information.

Inspecting during or after manufacture—This step involves the routine measurement of daily production at various phases of the process. This information is used for documentation purposes and for the next QC step (appraising conformance), in which decisions on product disposition are made. A successful QC program develops and implements the most "effective" measurement techniques for inspecting routine production.

Appraising conformance—This QC step compares the product information obtained through the measurement techniques, during or after manufacture, to established specifications. This step involves using a variety of statistical techniques that allow for the analysis of the collected product data and the comparison of these data to set specifications.

Planning for improvements—This QC step involves all components of a program, the measurement techniques discussed in detail in this section, as well as the planning and establishment of improved manufacturing practices, storage, distribution, and other practices within the manufacturing cycle of a product. In terms of the measurement techniques, this step involves:

- Calibrating the existing measurement techniques;
- Studying and testing alternative measurement techniques that provide improvements in terms of speed, accuracy, reproducibility, simplicity, and costs; and

- Using additional measurements at various production phases (e.g., in-process measurements).

    Sensory methods, as with all the other QC measurement techniques, play a role in each of the four steps above. Therefore, this book presents procedures to:

- Set *sensory* specifications;
- Train and operate a sensory panel (i.e., to establish the sensory measurement technique used to inspect the product's sensory characteristics);
- Develop a system to appraise conformance based on sensory characteristics; and
- Develop a program to monitor and improve panel performance (i.e., the sensory measurement technique).

MEASUREMENT TECHNIQUES IN A QC PROGRAM (INSTRUMENTAL AND SENSORY TEST METHODS)

The previous section outlined the steps in a QC program, in which product measurements play an important role. A detailed discussion of QC measurements is warranted.

Except for one of the five steps in a QC program—the actual manufacture of the product—all other steps are either (a) measurement techniques themselves or (b) related to and depend upon the product measurements obtained. Ultimately, the decisions on product disposition (reject, accept, hold) are based on the results of the product measurements themselves. This implies that the effectiveness of a QC program depends, to a large extent, on the type of measurement techniques used, their validity, reliability, and reproducibility. Therefore, one of the important functions of a QC department is to research, establish, and maintain the most appropriate measurement techniques. Often, measurement techniques are established and seldom reviewed or improved. Levy (1983) indicates that quality control programs often are involved in testing for the sake of testing, rather than understanding what the results actually mean. There is also a tendency to regard a test as sacrosanct, not to be changed or modified in any way.

The measurement techniques used at the plant level can be instrumental and/or sensory methods. Heretofore, focus has been primarily on instrumental methods. However, as consumer products companies realize that sensory testing is a critical part of product development and quality control (Pangborn and Durkley 1964) and recognize that sensory evaluation is developing as a science (Pangborn 1964), sensory methods occupy the same important role as instrumental tests.

Instrumental Methods in a QC Operation

Established QC programs in the consumer products industry have instrumental tests incorporated into their routine production measurements (Baker 1968; Langenau 1968; Finney 1969; Kramer and Twigg 1970; List 1984; Stevenson et al. 1984; Herschdoerfer 1986; Hunter 1987; Huntoon and Scharpf 1987). For certain products, the instrumental methods are the only ones needed and, indeed, are the most appropriate techniques to use. Products that fall into this category are automobiles, mechanical devices, and laboratory equipment. Control of these products is based exclusively on measuring very precise and analytical characteristics, such as dimensions, strength, durability, precise chemical composition, various performance characteristics, and so forth. In these examples, instrumental methods meet all the requirements needed for effective QC measurements, such as precision, accuracy, speed, and simplicity.

On the other hand, in the case of consumer products companies where products are manufactured to deliver certain sensory characteristics and performance to consumers, other test methods, such as sensory methods, might be beneficial and sometimes imperative in a QC program. Those companies ought to measure and control the quality of their products, based on what consumers directly assess: the product's sensory attributes.

Advantages of Instrumental Methods in a QC Program
Instrumental test methods present a variety of advantages that make them the ideal and ultimate choice in a QC program, to measure raw ingredients/materials and finished products, as well as to be used in in-process controls. These advantages are (Langenau 1968; Kramer and Twigg 1970; Szczesniak 1973; Levitt 1974; Noble 1975; Szczesniak 1987):

- Simplicity;
- Expediency;
- Immediate or quick turn-around of data;
- Continued operation (no restriction on number of products/samples tested);
- Precision;
- Accuracy;
- Reproducibility;
- Cost efficiency; and
- Compatibility with other instruments and/or computers.

Disadvantages of Instrumental Methods in a QC Program
The limitations of instrumental methods are well recognized (Friedman, Whitney, and Szczesniak 1963; MacKinney, Little, and Brinner 1966; Baker 1968; Sjostrom

1968; Bourne 1977; Williams 1978; Bruhn 1979; Trant, Pangborn, and Little 1981; Szczesniak 1987). Particularly relevant are:

- For some characteristics, the lack of instrumental techniques that measure properties related to the attributes of interest;
- In some cases, the lack of a relationship between sensory and instrumental measurements, as well as the inability of those instruments to relate and predict a sensory response;
- The inability of instruments to measure all the components of the total sensory response; and
- The lower sensitivity of some analytical methods or instruments (e.g., chromatography), compared with human sensitivity.

Assessment of Instrumental Advantages
and Disadvantages
The value of a new sensory program added to the current quality control capability requires the instrumental techniques in operation to be assessed. Additional measurement techniques may be needed; however, they will not be seen as necessary unless the instrumental information is examined. In the context of sensory evaluation, the authors have encountered many instances, in production facilities of consumer products companies, where instrumental tests are employed without the relationship between sensory and instrumental results being assessed. Often times, the results of studies in the literature on sensory/instrumental relationships are taken at face value and adopted by a company, without any further testing. Alternatively, some people believe that measureing physical or chemical properties is sufficient, since they ought to be related to the perception of a sensory attribute. Therefore, it is believed that measuring the physical or chemical property translates directly into perception measurements. Examples include the measurement of sodium chloride to predict perceived saltiness, the measurement of physical properties (e.g., particle size, stress/strain) to predict texture attributes, the measurement of reflectance to predict color, and so forth.

To assess if the instrumental techniques are sufficient in an established QC program, the following questions need to be addressed:

1. Are the product's sensory characteristics a criterion for consumer product selection?
2. Has the company studied the relationship between the instrumental and the sensory responses?
3. Do the instrumental tests in operation relate to all of the critical sensory characteristics?
4. Ultimately, are all critical sensory characteristics of the product being measured by the existing instrumental tests?

In case of a negative response for questions 2 through 4, either a sensory/instrumental study is needed and/or the company ought to consider implementing sensory tests, in addition to the instrumental tests in operation. Once a sensory/instrumental study has been completed, the authors' recommendations are:

- Use an instrumental/analytical technique in the case where a relationship between both methods exists. As an example, the Coca-Cola Company continues to conduct flavor evaluations, but is slowly moving to replace them with continuous instrumental monitoring of chemical and physical parameters (Woollen 1988).
- Consider implementing sensory methods, in addition to using instrumental methods in a QC program, when no relationship exists between both methods. Sensory methods are not substitutes for instrumental methods and need to be added to the program if they are the only tests that can measure critical product properties.

Sensory Methods in a QC Operation

Consumer acceptance of consumer products is affected by a variety of characteristics. Among them are their functionality, sensory characteristics, convenience, safety, cost, and so on. For many of these products, sensory characteristics, such as flavor, fragrance, and skinfeel properties, play an important role in their acceptance. The quality control program for such products is enhanced by the implementation of sensory methods, to measure and control the sensory properties of their products (Gerigk et al. 1986). Experts that have dealt with the relationship of sensory and instrumental data state that color, flavor, and texture measurements fall in the realm of sensory perception, and we must accept that they can only be measured directly in psychological and perceptual terms. Instrumental devices measure only physical properties that can be related to sensory properties (Szczesniak 1973; Noble 1975; Little 1976).

Advantages of Sensory Methods in a QC Program

Sensory methods have the following characteristics that make them unique in a QC program. Sensory methods:

- Are the only methods that give direct measurements of perceived attributes;
- Provide information that assist in better understanding consumer responses;
- Can measure interactive perceived effects; and
- Can provide "integrated" measurements by people versus "discrete" measurements provided by instruments.

A brief explanation of these characteristics follows.

Direct Measurement of Sensory Attributes

A sensory panel is the only "instrument" that can provide a characterization of the perceivable attributes of a product (Larmond 1976). Sensory information is very useful to companies, mainly because it provides a more realistic picture of what consumers themselves perceive and is often the only measurement of a given product property (e.g., behavior of a food product while being chewed).

The first aspect implies that the sensory measurements help understand what consumers perceive and like, while instrumental data often cannot. The second aspect implies that sensory methods provide an additional piece of information on attributes not easily or directly quantified by instruments. Some examples are the absorbency and moistness/wetness of lotions, the melting properties, oiliness, and perception of particles in ice creams (attributes related to the consumer perception of "creaminess"), flavor release in chewing gum.

Measurements Related to Consumer Responses

A consumer response is a two-component response. One component is the reaction to the "perceived" sensory attributes (e.g., saltiness, shininess, thickness, oiliness), and the second component is a personal/subjective reaction to those perceived sensory attributes. These subjective reactions are affected by personal preferences, expectations, past experiences, and other product parameters (e.g., cost, convenience). A consumer response then is a complex reaction affected by both perceptions and expectations. Analytical sensory methods, such as those implemented in a QC/sensory program, provide information on one of the two components of a consumer response: the perception of sensory attributes.

The direct relationship between analytical sensory methods and consumer responses offers two main advantages in the operation of a QC program: distribution of quality products and direct responses to consumer complaints.

1. Distribution of quality products
   Fewer defective (disliked) products are shipped to consumers, when a company determines the relationship between the consumer likes and dislikes (consumer responses) and the product's sensory attributes (sensory panel responses). This information, combined with a panel that measures the product's sensory properties, allows a company to ship products that have the sensory characteristics that yield high consumer liking. Each of the QC/sensory methods presented in this book addresses the research that investigates the relationship between consumer and sensory responses and presents the procedures for these studies.

2. Direct action on consumer complaints
   Because of the close relationship between consumer responses and sensory

measurements, a sensory program in a QC function allows a direct response to and solution for consumer complaints. Consumer complaints can be easily tracked to their source when sensory characteristics are measured at the plant. For example, a consumer complaint, such as "bad odor" in a product (e.g., soap bar), can be tracked back to the sensory measurement collected on that production batch. If a sensory attribute, such as animal/hide, is noted at relatively high intensities, it indicates the need to more carefully control the level of raw materials associated with this off-note, since it elicited consumer complaints. Furthermore, the description of these off-notes may direct the QC and processing staff to the possible sources of these off-notes to solve this problem.

Measurement of Interactive Effects

Sensory methods can measure the interactive effects of attributes or ingredients in a product. People perceive products as a "whole" and, therefore, can measure the perceived interactive effects of stimuli (e.g., ingredients), which are often nonadditive. A sensory panel is the only "instrument" capable of measuring these effects because instrumental techniques can only measure the concentration or amount of the individual components. The analytical measurement of each component cannot always be added, to predict the perceived effect. A simple example of this phenomenon is the interactive effect of basic tastes, such as sourness/sweetness, sweetness/bitterness, and so forth. While an instrument may provide identical readings for the acid concentration in an acid solution and in an acid/sucrose solution, a sensory panel's rating for perceived sourness is different for both solutions. In the acid/sucrose solution, the perceived sourness may be lower, due to the interaction between both chemicals.

Measurement of "Integrated" Product Responses

Sensory responses that encompass more than one primary attribute (therefore, "integrated") are often important quality control measurements. Among these are product assessments, such as overall difference, overall quality, acceptance/rejection responses, and in/out measurements. In these cases, panelists integrate the perception of all product characteristics, to provide one final product measurement. This product measurement is the basis for decisions about product disposition.

While "integrated" responses can be provided by humans (who can integrate), instruments (which are separators) provide discrete and unidimensional measurements (Pangborn 1987). Although sensory methods should also collect discrete and unidimensional results (e.g., the intensity of singular attributes, such as color intensity, fruitiness), collecting integrated responses can be beneficial. Implementing a sensory program as part of a QC function provides that opportunity.

Disadvantages of Sensory Methods in a QC Program

The disadvantages of sensory methods in a QC program are the same as those of other sensory programs (Dawson and Harris 1951; Amerine, Pangborn, and Roessler 1965; Foster 1968; Kramer and Twigg 1970; Levitt 1974; Trant, Pangborn, and Little 1981; Stone and Sidel 1985; Meilgaard, Civille, and Carr 1987; Pangborn 1987). Among those are:

- Time involved;
- Expenses incurred;
- Effect of environmental and emotional factors on people's responses;
- Effect of biases and physiological and psychological factors on the product's results;
- Possible lack of accuracy and precision;
- Attrition; and
- Need to handle and resolve personal factors that affect the operation of the program (e.g., sickness, layoffs, promotions, moves).

When sensory methods add the unique measurement "to the set of QC measurement techniques" (e.g., mechanical, microbiological), the above disadvantages are accepted in order to obtain a more complete picture of the product's quality.

THE HISTORY AND PRACTICES OF QC/SENSORY PROGRAMS

History and Developments

The history of QC/sensory programs has undergone two distinct phases. The first and oldest QC/sensory practices date back to the development of sensory evaluation methods for product evaluation in industry. These programs relied on "experts," such as perfumers, brewmasters, or winemakers. Operating such a "QC/sensory program" was the sole responsibility of such experts, specifically by establishing the criteria to judge quality. The actual judgement of quality was done on a subjective basis (Dove 1947; Hinreiner 1956). In time, this responsibility was delegated to more "experts" or to a small panel of judges. The sensory practices used at that time mainly included evaluating the product's quality (e.g., poor to excellent) and, to some degree, evaluating product characteristics by ranking or rating (Platt 1931; Plank 1948; Dawson and Harris 1951; Peryam and Shapiro 1955; Amerine, Roessler, and Filipello 1959; Tompkins and Pratt 1959; Ough and Baker 1961).

An important turning point in quality and quality control judgements occurred when attention was given to selecting and training judges, selecting a formal sensory methodology (including using reference materials), and using formal data analysis methods (Dawson and Harris 1951; Pangborn and Durkley 1964). QC/sensory practices continued to be developed, such that, in the 1960s and 1970s formal sensory evaluation programs were developed and integrated into corporate quality control functions. Regrettably, these few companies have implemented methodology that has yet to be published. It was not until the end of the 1970s and beginning of the 1980s that some of the industrial QC/sensory practices or concepts were discussed within the sensory community.

Two IFT (Institute of Food Technologists) Sensory Evaluation Division Symposia, "The Role of Sensory Evaluation in Product Quality Assurance" (1979) and "The Wide Scope of Sensory Evaluation" (1980), addressed the industrial practices of sensory evaluation in a quality control/quality assurance function, for the first time. In these two symposia, four of the papers (Nakayama and Wessman 1979; Wolfe 1979; Reece 1979; Merolli 1980) discussed very relevant, applied industrial practices of and issues in a QC/sensory program, such as:

- Requirements of the program: plant testing facility and personnel (sensory coordinator and panelists);
- Applications of sensory methods for controlling ingredients, in-process materials, packaging, and finished products;
- General procedures for training a QC panel; and
- Applications and problems associated with the use of reference standards in sensory measurements.

These papers are very general in nature and do not present specific technical information, discussion on methodology, or procedures in the program implementation. However, they are the first documentation of the philosophical, administrative, and general technical issues involved in an in-plant sensory program.

In 1985, an IFT symposium entitled "In-Plant Sensory Evaluation" addressed three areas in the operation of QC/sensory evaluation programs:

1. Initiating QC/sensory programs (Rutenbeck 1985);
2. Implementing and operating such programs in various industrial situations: in a small processing operation (Mastrian 1985), in multiple plants (Stouffer 1985), and in multiple plants with an international program (Carlton 1985), and;
3. Computerizing data entry for those programs (Powers and Rao 1985).

These papers are also general in nature, in that specific QC/sensory procedures or product examples are not discussed. The value of this literature is that, in contrast to the few early publications (1979–1980), some papers of this 1985 symposium address real and important situations and problems of QC/sensory programs not published before. For example, Rutenbeck (1985) discussed two important steps required prior to implementing a QC/sensory program: assessing the need for such a program and selling the program. In assessing the program needs, the following issues should be addressed, according to Rutenbeck:

• Objectively reviewing the product history (consumer complaints, product holds, and rework), to determine if the QC/sensory program is needed;
• Assessing the current monitoring device, to determine its effectiveness;
• Assessing additional monitoring techniques, if needed—physical, microbiological, chemical, or sensory; and
• Identifying the products, processing phases, and attributes, to be evaluated in the sensory program.

The problems discussed in selling this program to management are critical and are faced by all sensory professionals who plan to get a QC/sensory program started. Rutenbeck (1985) discusses what getting management support means, as well as some activities needed to get this support. These activities include:

• Demonstrating the benefits of the program to the plant and corporation;
• Cultivating interest among non-sensory groups;
• Calculating overall program costs;
• Scheduling multiple presentations at all levels of the corporation, to sell the program; and
• Discussing the return on management's investment, in terms of product consistency, reducing consumer complaints and product reworked, and increasing sales volume.

Stouffer (1985) discusses important aspects involved in training panels at different sites. Among these are the role of the corporate (central) and on-site coordinators, interacting with local plant management, developing uniform practices across plants, reviewing the multiple plant data, and attaining consistent results from all plant panels. Carlton (1985) discusses additional aspects to consider when developing QC/sensory programs at the international level. Cultural, language, and social differences also have to be considered.

There is not, however, a publication that approaches the subject of QC/sensory in a comprehensive, practical manner. An ASTM publication—*Standard Guide for the Sensory Evaluation Function within a Manufacturing Quality Assurance/Quality Control Program*—currently being developed by the members of the subcommittee E-18.07, is the first publication that reviews various sensory methods used in a QC operation and discusses some of the characteristics of QC/sensory programs. This book is developed to provide the quality and sensory professionals with a step-by-step description of the procedures for implementing one of several QC/sensory programs that meet the product and company needs.

Published Methods Used in Quality Control/Sensory Evaluations

The published QC/sensory evaluation literature only covers the sensory tests used for quality control measurements, without a detailed explanation of the aspects involved in establishing a program. These articles can be divided into two categories. One group makes reference only to the type of tests used (Rutenbeck 1985; Dziezak 1987; Chambers 1990; Hubbard 1990). The other group presents a more detailed description of the type of test or scale used (Bruhn 1979; Ke, Burns, and Woyewoda 1984; Bodyfelt, Tobias, and Trout 1988). None of these publications discusses advantages or disadvantages of the recommended method. Furthermore, no alternatives to the recommended method are discussed.

Eight types of sensory tests for quality control purposes have been cited in the literature:

1. Overall difference tests (Kramer and Twigg 1970; Mastrian 1985; Dziezak 1987; Chambers 1989);
2. Difference from control (Aust et al. 1985);
3. Attribute or descriptive tests (Rutenbeck 1985; Chambers 1990);
4. In/out of specifications (Nakayama and Wessman 1979; Sidel, Stone, and Bloomquist 1983);
5. Preference and other consumer tests (Kramer and Twigg 1970; Dziezak 1987);
6. Typical measurements (Steber 1985);
7. Qualitative description of typical production (Ruehrmund 1985); and
8. Quality gradings (Kramer and Twigg 1970; Bruhn 1979; Waltking 1982; Ke, Burns, and Woyewoda 1984; Rothwell 1986; Bodyfelt, Tobias, and Trout 1988).

Three of the eight methods—attribute/descriptive, in/out, and difference from control measurements—are the most sound sensory tests for quality control purposes. The three methods are presented in Chapters 3, 5, and 6. The other five methods are not recommended in this book, based on certain limitations.

Difference Tests
Forced choice difference tests are too sensitive and too general for quality control purposes. In an overall difference test, judges focus on any difference between products, small or large. Unless each production batch is identical to the control or

standard, large amounts of the production will be found to be "significantly different" and consequently rejected. No measure on how different the product is from the control is obtained.

The method of difference from control (Aust et al. 1985) was developed because of the limitations of overall difference tests. Therefore, this approach is the alternative to forced choice difference tests in a quality control setup. Overall or forced choice difference tests, however, are very sound sensory tests that have many other appropriate applications in research and development.

Preference Tests

Preference and other consumer tests are the least recommended sensory methods for quality control measurements. Using preference tests is considered inappropriate because of the population (participants) used and the type of data collected. The participation of a small group of company employees is considered unrepresentative of the true consumers and is therefore of little value. A production batch preferred by employees may not be the one preferred by the actual product consumers.

Like difference tests, preference tests provide no information as to how or to what degree the production sample differs from the standard. Furthermore, a product that is equally preferred may have distinctly different sensory properties than the control. Also, decisions based on "significant preference" for the control/standard over production would mean rejecting a large percentage of daily production.

Typical Measurements

Determining "typical/atypical" production (Steber 1985) is one of the most popular QC product evaluation methods used in consumer products companies. Its popularity is due to its simplicity and the "directness" of the results. Production is classified into "typical" and "atypical." If production is "typical," it is shipped, and if it is "atypical," it is held.

The two main disadvantages of the method are the uninformative nature of the information and the subjectivity of the evaluations. These measurements are considered general and nonactionable, when production is found "atypical." No information on the nature or magnitude is obtained. On the other hand, panelists are subjective in that they focus on different sensory attributes, to judge what represents typical and atypical production, and have different tolerance criteria for variation.

Qualitative Assessment of Typical Product

This procedure uses a qualitative description of typical or high-quality production as a base to judge daily production. The product is considered unacceptable if it does not match this qualitative description. Ruehrmund (1985) and Spencer (1977) show this type of evaluation for the quality control of coconut and ice cream, respectively. The criteria to judge the smell of coconut (Ruehrmund 1985) are: "clean, fresh—no stale, musty, or other foreign odors." The descriptions for unacceptable feel are: "dry and free flowing—no damp or greasy feeling." This type of evaluation has two major problems. First, the descriptors (criteria for judgement) are too vague and subjective for panelists to know and agree on the characteristics being evaluated. This applies to the criteria of "clean and fresh," given by Ruehrmund, and "freshness/cleanness/distinctiveness," given by Spencer (1977). The criteria of this method have been found to be too general and "integrated" (i.e., encompass the evaluation of many attributes) to be useful for management's decisions (Bauman and Taubert 1984). These terms are classified under "consumer terms" and are appropriate for consumer tests, but not recommended for any analytical sensory test at the R&D or plant level. The second disadvantage of this method is the lack of information and guidance when the product is assessed as "atypical."

Quality Ratings
This approach, together with the "in/out" and "typical/atypical" evaluations, has been the most popular QC/sensory test used. Furthermore, the quality gradings have not only been used in in-plant situations, but in academia, research, and government evaluations. Various trade associations and commodity organizations have developed and published "standard" procedures that fall into this category. Of special interest are the dairy and oil standard grading systems developed by the American Dairy Science Association and the American Oil Chemist Society (AOCS). These grading systems or scorecards consist of any combination of these components:

- A grading (e.g., A, B);
- A scoring system with points assigned for each grade;
- An overall quality grade (e.g., excellent to unacceptable); and
- A description of sensory characteristics (also referred to as references or standards).

Quality grading systems developed for various commodities are those for eggs (Institute of America Poultry Industries 1962), oils (Waltking 1982), ice cream (Spencer 1977), dairy products (Bodyfelt et al. 1988), and squid (Ke et al. 1984).
Despite their popularity and extensive use, these quality grading systems have serious limitations and flaws. Among them are:

- Using vague, integrated, and technically incorrect sensory terms;
- Clustering or grouping several distinctly different sensory terms into one quality category/grade;

- Lacking attention to all the sensory attributes that affect the product's acceptance;
- Lacking equidistant quality grade points or categories;
- Nonuniform assignment of score weights across quality categories;
- Using qualitative and quantitative factors in a single category grade; and
- Inappropriately using statistical analysis of data collected, using scales with characteristics described above.

Several authors have evaluated these limited scoring systems and presented their criticism (Sidel, Stone, and Bloomquist 1981; O'Mahoney 1979).

Improved quality scoring systems are currently being developed. Among these are the procedures documented by the ASTM task group E18.06.03 on edible oils and the new scoring system developed and published by the American Oil Chemist Society (AOCS 1989).

2

Program Design and Initiation

PRELIMINARY DETERMINATION OF PROGRAM PARAMETERS

In any sensory study, the objective of the sensory test or program is based on the project situation or problem. In a QC program, defining the sensory project objectives requires one to consider several aspects, such as product variability, marketing objectives, and manufacturing conditions. These considerations have an impact on important decisions of the QC/sensory program, such as selecting the product category, points within the process at which to evaluate, the plants to be included, the key sensory attributes to be evaluated, the product sampling, and the test method to be used.

Companies cannot hope to understand consumer acceptance, consumer complaints, or shifts in market share, without monitoring the product's sensory properties that drive these consumer responses. The key sensory test objective of a QC/sensory program is to develop an efficient way to understand the key attributes that affect consumer liking or acceptance, to assess which raw materials and processing variables affect the final sensory properties, and to develop the most efficient and timely system to measure and control these sensory attributes that are so important to consumer acceptance.

One critical marketing decision that must be made within the company when establishing a QC/sensory program is whether the selected product is expected to be consistent across the geographic area in which it is marketed. This may be a regional (southwest United States) national (all of the United States), multinational (all of Europe), or global product. A key question to be determined by marketing and upper management is whether the product is expected to be the same, in terms of sensory properties, in all markets or different ones, depending on the market.

Until this critical decision about deliberate consistency or variability of product by location is made, no QC/sensory program can be implemented to measure and control variability.

Identification of Products to be Included in the Program

Companies that produce a variety of products need to decide which products in which product categories are to be included in a QC/sensory program and in which order each product is to be introduced into the system. Within a company's general QC program, all products are tested for one or more physical or chemical properties, as a test of product consistency. Similarly, any QC/sensory program should encompass and evaluate *every* product. In the initial design of a QC/sensory program, it may be desirable to include all products, but it may also be impractical because of the requirements in terms of time, money, and personnel to do all products simultaneously. Therefore, one or a few products need to be selected as the "first" or "test case" in the program. This product selection requires careful consideration of several criteria, by which to determine the best choice for the start of the program. Some criteria to be considered are market position, critical problems, production variability, and multiple plants.

Market Position

The products that yield the largest profits for a company are those usually included first in the system. Position in the market place implies high consumer acceptance of the product's sensory attributes and requires assessment and control of these products and their sensory attributes.

Critical Problems

A sensory defect in a product that is considered unacceptable by management or the product's consumers (such as the presence of an off-flavor or off-color and its persistence across several production days) is the most important reason for including a product in a QC/sensory program. This decision is especially critical, if the problem has already begun to affect consumer acceptance, demonstrated by consumer complaints, loss in shares, or poor performance in consumer tests. The negative consequences of shipping product with a sensory problem to loyal consumers, because of an inability to detect and control such problems, strongly justifies measuring and controlling that product's sensory attributes. Sensory measurement may be the only way to determine a product's critical problems, as well as the level of variability in production.

Production Variability

Consumers want products not only high in quality and low in defects, but also with a consistent quality. Therefore, products with large production variability should be considered as candidates for the QC/sensory program because of the potential negative impact on consumer acceptance. A program, which can measure and control production variability, should include those products with large production variability. With many products, instrumental methods may not be able to measure or track product variability.

Multiple Plants

Although a lesser consideration, companies with more than one production facility may choose to include a product that has met one or more of the above criteria and is produced at more than one plant. This permits study of quality and consistency across plants.

Points of Evaluation Within the Process

Early in the program, a decision must be made concerning which stages of the process require evaluation of samples. Finished products are of primary interest in a QC/sensory program, since these are the products that the consumer sees and, thus, accepts or rejects. However, earlier in the process, both ingredients and in-process materials can be evaluated as part of a total QC/sensory program.

Raw Ingredients/Materials

Once an inferior or defective raw material enters a process, the defect may be magnified several times, as the defect is incorporated into and thus becomes detectable in the finished product. One very deliberate and effective way to control the quality of a product is to control the quality of the materials used in producing that product. Therefore, evaluating and controlling the incoming raw materials' sensory properties offer a significant opportunity to monitor and eliminate variation early in the process, thus saving time, money, and product. When possible, every effort should be made to develop a QC/sensory effort to evaluate all or most raw ingredients by 1) setting ingredient sensory specifications, 2) conducting regular evaluations upon receipt at the plant, and 3) making decisions as to the disposition or blending of inferior raw ingredients.

Selecting the raw ingredients to be monitored is determined by the finished products to be controlled and by the critical sensory attributes of those finished products known to impact on consumer acceptance. It is possible to incorporate the assessment of many raw ingredients into a program, since the analysis of raw ingredients generally consists of the evaluation of only one or two attributes as opposed to five to ten attributes, in most finished products. One raw ingredient may be incorporated into several finished products. Therefore, the control of only a few raw ingredients may influence the control of several finished products. In contrast, selecting in-process materials and finished products for a QC/sensory program is more complex.

Understanding the relationship between raw ingredient sensory attributes and the sensory properties of the resulting finished products allows for the eventual shift from the more complex multi-attribute evaluation of several finished products to the simpler, more efficient evaluation of a few key attributes of a few key raw ingredients. This system insures a quality product through early control, rather than through large scale disposition or rework of finished product.

In Process

For some processes, there are critical points at which sensory tests can be done to measure and control the introduction of sensory attributes into the product. Just as analytical chemical and physical tests are conducted at stages throughout a process, QC/sensory monitoring is recommended at critical stages in production, where interim product quality reflects the quality of finished product. When a batched lot of one phase of a product, such as a dough premix for a cookie or a fragrance blend with carrier for a household product, are introduced into the process, this in-process phase or batch can be tested for key sensory attributes. Often, the reaction of raw materials to partial processing reveals characteristics that are not desirable in the finished product. As with raw materials, if defective or variable in-process products can be identified and held before final processing, conversion, or inclusion with other in-process materials, the result is the saving of time, production costs, and product.

Finished Product

Since almost all consumer products are purchased for their sensory properties, as well as other performance characteristics, it behooves managements of consumer products companies to monitor the sensory properties of their products before they are shipped to the marketplace. In spite of an effort to control the sensory quality of a product by monitoring and controlling raw materials, it is, at least initially, necessary to evaluate the sensory characteristics of finished product. The interaction of raw materials and/or the effects of processing on the already cleared raw materials requires monitoring. The appearance, fragrance, or flavor and texture properties can be monitored for some time period, to learn the variability per se and its relationship to any variability in the raw materials. Once management and the QC/sensory team understand how raw materials affect finished product, it is possible to reduce the frequency of evaluation of the finished product.

Quality control of consumer products often includes the instrumental analysis of chemical and/or physical properties at different processing stages, but often does not include monitoring appearance, aroma, flavor, or texture of these same products. In many cases, quality control sensory measurements early in the process can also reduce the need for frequent sampling of finished product. This is especially true when the effects of early detection of problems have been related to problems in the finished product.

Sensory Attributes Evaluated in a Product

Two major factors determine which sensory attributes are to be evaluated in a product: attributes with variability and attributes that drive consumer acceptance. These criteria are relevant to those attributes that define raw materials, in-process product, or finished product. Only the sensory attributes of finished product can be directly evaluated by consumers. However, the effects of variation in the sensory properties of raw materials and in-process product can be related to the sensory attributes found in finished product.

Attributes with Variability

Product variability is a major reason for a product to be included in a QC/sensory program. The product variability in sensory terms is measured as variability in sensory attributes. Generally, sensory attribute variability can result from the presence of an "off-note" in a product, or from the variation in intensity or strength of sensory attributes that are normally found in the product. An off-note is a sensory attribute (plastic flavor, bitter taste, gritty texture, a color) that is not present in the product under the best or optimum conditions. The particular attribute is considered "atypical" of the product profile and may be the reason for a rash of consumer complaints.

The shift in level or intensity of the characteristics typical of the product is a potential source of product variability. Although a product may possess the proper set of attributes, one or more may be shifted out of the normal intensity range (too sweet, not crisp enough, too dark, not shiny enough). In these cases, any one attribute may be too high or low, or more than one attribute may be outside what is typical for regular production or a designated control. Therefore, the attributes to be included are those that vary outside the normal range for the typical product and those that are "off-notes" or not normal for the product.

Attributes that Drive Consumer Acceptance

One of the most surprising aspects of measuring both sensory attribute variability and its impact on consumer acceptance or liking is that the size of the variability in an attribute is *not* necessarily related to the size of its effect on consumer responses.

When consumers see a wide range of color intensity in cookies or potato chips in one package, they may respond negatively only to the extremes, and may rate a very wide range of color as highly acceptable. Consumers may also find a wide range of fragrance levels acceptable in personal products. Conversely, however, a low-level indication of staleness or rancidity in the cookies or potato chips may yield dramatic loss of consumer acceptance. Likewise, a slight grayness in the color of a shampoo may reduce overall acceptance, as well as acceptance for appearance or cleaning ability.

The criteria for including sensory attributes into the QC/sensory program are 1) those that induce negative reactions in consumers and loss of consumer acceptance and 2) those that management feels are too broad a range to represent a clear concept of the product.

Chapters 3 to 6 cover the different methods that can be used in a QC/sensory program and include a discussion of attribute selection for a variety of consumer products, as well as the criteria used in attribute selection.

Sampling

During the phase of identification of the parameters to be included in the QC/sensory program, it is necessary to determine the amount of sampling that will have to be done to understand the variability across products, attributes, and plants. Preliminary information may be available that indicates certain plants vary more than others because of the processing equipment or ingredient sources. Certain key attributes may be a problem only on certain shifts or specific times of a start-up on a production line.

Early assessment of the scope and frequency of product variability provides a clue to the possible sampling schedule needed during the preliminary implementation phase. It is from careful evaluation of possible sources of variability that an effective sampling plan can be developed, to cover variability across plants, products, and attributes. Journals such as *Technometrics* and *Journal of Quality Technology* present technical approaches for determining the best sampling plans for a wide variety of product variability interactions.

IDENTIFYING RESOURCES

After the characteristics of the QC/sensory program are determined, the next step, before the design and implementation phase, is assessing all possible resources that are available to support the program. The areas to be studied include the methodology, the personnel, the physical facilities, management support, and the availability of outside resources.

Methodology

Recommended Methods
Within sensory evaluation, there are several sensory methods to meet project objectives in basic research, product development and improvement, product maintenance, and marketing/marketing research. Given the nature of manufacturing and quality control requirements, the sensory evaluation methods that are best suited to meet those QC needs are those that measure variability. These methods include those that utilize ratings and those that assess overall conformance to a product concept.

Rating Methods
Methods that involve ratings of 1) attribute intensities, 2) quality, or 3) difference afford the sensory analyst the capability of determining a *degree* of variability from some target control or specification limit. The three methods prescribed in Chapters 3, 4, and 6 fall into this category. Each is described below with general application, advantages, and disadvantages.

   The Comprehensive Descriptive Method.   This method involves the rating of intensity of individual key attributes, for which specifications are set. This method identifies the samples that are out of spec and to what degree for each attribute.

Application: For a company's major flagship brand, this method careful assesses product variability, in terms of both key attributes and the direction they may vary. Such in-depth information provides direction in making changes in ingredients or process to increase greater product conformance.

Advantages: This method provides the most detailed information about "what" is varying in a product, the size of that variability, and the direction (too high or too low), relative to the product specifications. This method most closely resembles the evaluation of other physical or chemical characteristics that are integral parts of a broad scope QC program.

Disadvantages: This approach is costly and time consuming, mainly for two reasons: the sensory tests required (e.g., descriptive and consumer research tests) and the time and resources needed to establish and maintain the program (e.g., extensive training samples and references needed).

   The Degree of Difference/Difference from Control Method.   This method involves rating each sample in terms of the *size* of the difference from a designated control. A cut off point is designated to assess which products or ingredients are too different from the control to be within specification limits.

Application: When product or ingredient variability follows a single continuum, from the control to less desirable or "off" production samples, this method provides a simple, one-step, method for assessing out of spec samples.

Advantages: This method requires few resources and little time to implement. Only one set of training references is needed, to establish a range of products from the control (not different to those very different from the control). Panelists need to learn to use only one continuum by which to judge sample conformance.

Disadvantages: A difference from control rating only defines the size of the difference from the target control but fails to provide information on the reasons for the difference. To establish the cutoff for "too different," some consumer research is required.

The Quality Rating Method.   This method involves rating each sample and/or some set of key attributes for quality. A specification is established for each attribute scale, to determine proper disposition of the tested sample or product.

Application: In industries or corporations where quality of a product or ingredient is commonly understood, developing quality criteria overall and for specific attributes can be done quite easily and can be used effectively.

Advantages: A quality rating program has broad history and understanding in consumer products, especially among food commodities. Therefore, the data are easily understood and interpreted by management.

Disadvantages: In cases where the notion of "quality" of a product is subject to individual personal preferences, the vagaries of shifting product concept, or the competitive set, quality ratings are not recommended. The shifting criteria create an inconsistent rating system. In addition, the training can be very complex, if common quality criteria are developed and used.

Nominal Methods

Methods that allow trained panelists to decide if a given sample is within or outside of a product concept are also applicable in a QC/sensory program. In a QC application, the concept represents the range of attribute intensities considered "acceptable" or "in-spec."

The "in/out" method requires the initial careful definition of the characteristics of "in" or "acceptable" product. The subjects involved in product evaluation are shown several examples of the sample, a product, or raw ingredient, some that are within the product limits or acceptable concept ("in") and some that are outside the product limits or concept ("out"). During subsequent testing, these trained panelists are expected to determine if the test products or raw materials fall in or out of the designated "acceptable" product concept.

Application: The choice of the "in/out" method is appropriate for those situations in which 1) a simple yes/no answer provides sufficient information for

QC/sensory, 2) the capacity to identify and/or retain reference samples is limited, or 3) the resources to train and maintain one of the more complex rating methods are unavailable.

Advantages: The simplicity of having only one criterion, one evaluation, and one simple result is a significant advantage of the "in/out" method. Training panelists is relatively easy, and QC management knows immediately which samples to ship.

Disadvantages: Conversely, the simplicity reduces the amount of specific information about the reasons for a sample's "out" designation. Decisions on disposition require some further assessment of how "out" the sample is. Only then can efforts be made to correct the ingredient or process conditions that contributed to the product's "out" rating. A major disadvantage is that the decision process, normally left to QC, is part of the "in/out" method. Rather than providing evaluations and ratings that are used to decide on the acceptance or rejection of the product, judges using the "in/out" method decide directly yes/no or accept/reject, as part of the judgements.

Methods to Avoid

Some tests, such as most Overall Difference tests, are not effective for QC sensory work. Triangle, Duo-Trio, A/NOT A, and Simple Difference tests are very sensitive to fairly small differences. Therefore, by yielding only "different" or "not different" answers, they tend to lead to rejection of a large proportion of regular production through the assessment of only statistically significant differences. The *amount* of difference is often the key question in a QC environment, and difference testing alone will not determine the size of the difference.

Affective responses, including preference and acceptance, are not appropriate for routine production evaluation. The determination of the preferences or liking of a small group of plant personnel does not represent the population that routinely uses the product. However, it should be noted that some affective acceptance tests using a large consumer base from the population of users should be utilized in every QC program. The acceptance information, used in conjunction with other sensory tests at the early stages of a QC program, help determine how variation in certain attributes drive the acceptance of a product. This affective information is used to set the sensory specifications.

Personnel

The success of a QC/sensory program relies heavily on two categories of personnel: the sensory professional(s), who are responsible for program design and execution, and the sensory panelists, who are responsible for the routine evaluation of products and/or raw materials.

Sensory Staff

Designing, administering, and implementing the QC/sensory program require the expertise of different sensory professionals, depending on the size and scope of the program. The personnel requirements for a given program might include a sensory coordinator, a sensory professional at each plant, and one or more sensory technicians working with the plant sensory professional and R&D sensory coordinator.

The R&D sensory coordinator may be a member of the current QC staff who shows some interest in the sensory activity. In such a case, sensory training regarding methodology, panelist management and training, and test controls is critical. More often, the sensory coordinator comes from R&D and is given the responsibility for the QC/sensory program across several plants. This individual may have the sensory qualifications, but requires some grounding in plant and QC practices and in product ingredients and processing.

In most companies, the R&D sensory coordinator has a reporting relationship to both the R&D sensory manager and the QA/QC team or director.

The R&D sensory coordinator has the following responsibilities:

- Oversee the initial design of the QC/sensory program, by working with QC upper management.
- Define the scope of the program in terms of the products, attributes, processing steps, and sampling schedule to be included in the testing.
- Assess the companies current potential resources in terms of methodology, personnel, facilities, management commitment, and support functions.
- Design and develop the sensory program in conjunction with the quality control department, in terms of selecting and implementing the approach and preliminary stages of the QC/sensory evaluation program.
- Interview, select, and train plant sensory professionals, as each plant is brought into the system.
- Set criteria for screening and selecting panelists at each plant.
- Develop and conduct a training program to be implemented at each plant on the start-up of each plant program.
- Oversee the initiation of the established program after training.
- Be a resource consultant to all plant sensory analysts.
- Coordinate the QC/sensory program across all plants in terms of panelists performance, data handling, and reporting. In addition, oversee dissemination of the QC/sensory program results to QC and the other functions within the company.
  The plant sensory analyst has the following responsibilities:
- Assist the R&D sensory coordinator in implementating the preliminary stages of the QC/sensory program.
- Screen, select, and assist in training subjects as plant panelists to evaluate samples according to the designated program methodology.

- Manage samples, references, and controls (for product and raw materials) in terms of pickup, logging, and storage, in accordance with the program's prescribed sampling plans.
- Supervise the preparation and presentation of the samples for each evaluation session, using strict test protocols to minimize bias due to poor product handling.
- Coordinate panelists and the schedule of evaluation sessions. This includes working with panel supervisors to assure full and timely attendance.
- Maintain the technical skills and motivation of panelists.
- Manage the maintenance of the sensory facility and supplies.
- Manage sessions, to insure compliance to the test evaluation protocols for each panelist.
- Collect and compile data for each sample and report.
- Maintain interaction with R&D coordinator for technical issues and growth of the program.

One or more QC/sensory technicians may be required to assist the R&D sensory coordinator and each plant sensory analyst in the following responsibilities:

- Implement the routine sample handling.
- Conduct the routine sensory tests.
- Contact and schedule panelists and tests.
- Enter or record data from each test.

Panelists
As with any QC measurement, the "instrument" is a critical consideration. In sensory evaluation, the panelists are the measurement tool. Issues to consider when selecting, training, and maintaining panelists are as follows.

Selection
Panelists may be recruited from within the plant or R&D site where evaluations are to take place or from the neighboring community. Plant employee panelists may include line workers (when union and personnel limitations have been considered) or office staff, if use of line workers is limited or untenable. A major advantage of using plant personnel for QC/sensory evaluations is that a quality focus is communicated throughout the plant by these plant employee sensory panelists. However, using panelists from the neighboring community may be necessary, if personnel from within the plant are unavailable for the necessary regularly scheduled sample evaluations.

Sensory testing is labor intensive and using plant personnel requires adding personnel or extending work schedules to compensate for panelists' time for evaluations. Accommodations in staffing may have to be made for employee panelists who are absent from regular shift or office work for more than a 5 to 10 minute period.

Panelists' general requirements include general good health, interest in a sensory program long term, and availability on a regular basis (perhaps up to one-half hour every day). Each panelist is screened, using acuity tests that assess ability to discriminate and describe specific sensory properties. The actual tests and sensory properties that need to be considered are defined by the selected methodology.

Training

When panels are implemented across different plants, it is critical that the same selection criteria and training program be used at each plant, to insure that equivalent "sensory instruments" exist at each site.

The selected sensory program (see Chapters 3–6) defines the sensory method for which the panelists are trained. The simpler in/out or difference from control methods may require as little as 10 hours of training. The quality rating and comprehensive descriptive programs, which may involve rating several attributes, may require over 40 hours of training and practice before the panel is fully functional.

Maintenance

Panel motivation is secured by developing a reward system for panelists early. The size and frequency of the rewards are based on the skills and time provided by the panelists, the rewards permitted under union or nonunion personnel rules, and the commitment of the company and sensory analyst to recognizing a job well done.

Monitoring panelist performance is critical in all QC/sensory programs and will depend on criteria derived from the selected methodological approach. Each plant analyst and the R&D sensory coordinator must also develop a maintenance program that involves regular use of references and controls, regular feedback on performance, and remedial training programs for individual panelists or whole panels.

Management Support

From the very inception of a QC/sensory program, all sectors of the company's management need to understand, make contributions, and ultimately commit to the proposed program. One major responsibility of the quality team, charged with developing a QC/sensory program, is to determine both the commitment and contribution to be made by each sector of management.

A preliminary proposal, which outlines the program and its costs and benefits, should be presented to company management before any action is taken. Early decisions about the scope of the program, encompassing things such as the products or product lines, should be included in the proposal. Input from the various directors or vice presidents is used to modify the program.

Once management has had input into the program design, a more detailed proposal, which addresses the processing stages at which evaluation will be conducted, the sensory attributes to be measured and a tentative sampling schedule can be presented to management. In addition, the preliminary recommendations for the methods, facilities, personnel, and required outside resources are recommended. At this point, the various directors and vice presidents commit to providing information and re- sources, as described, in the following paragraphs, for each group.

Throughout the preliminary implementation stages, each director/VP commu- nicates support within his or her group so that the necessary consumer, sampling, processing, or physical data are available to the quality team. Once the consumer data and preliminary product information are collected and compiled, the recom- mendations, product attribute specifications, and program costs are presented to management to get final approval to move the QC/sensory program forward to full implementation.

Lack of full involvement of upper management and the resulting support from each group creates potential problems both in implementing and operating a QC/sensory program. Specifically:

- Without the input from the management of all groups, the key information on product, process, or consumer may be missing from the preliminary stage.
- Without commitment and support from the management of all groups, the necessary resources—time, personnel, facilities, product information, and so on—may not be complete. The success of the program, no matter how well planned and executed, may be at risk.

The responsibilities from each sector are defined broadly below.

1. Quality management needs to see the sensory aspects of the program as an integral and important part of the quality commitment throughout the com- pany. Since the sensory properties of most consumer products are the primary reasons for product acceptance and purchase, emphasis on sensory quality and its measurement are critical to the perception of overall product quality.
2. Research and development management needs to contribute insight on the types and size of product sensory variability, the ingredient and processing factors that influence variability, and consumer reactions to that variability. The R&D sensory input, in terms of personnel, training, expertise, and consul- tation, represents a major asset to the overall program. Understanding the technical aspects of a product's chemical and physical properties that relate to perceived sensory properties can be a major contribution. Understanding which attributes are critical, as well as the development of instrumental meth- ods to correlate with sensory measures and sensory specifications, provide key components to the program.

3. Marketing and marketing research management needs to understand the effect of quality on market shares and volume and generally participates significantly in the QC/sensory program. They provide information about consumer responses to product attributes, user demographics, and consumer complaints. These contribute to the early development of product attribute selection and specification limits.
4. Production or operations management, more than any other groups, first needs to understand the direct benefit of a well developed QC/sensory program. Since the manufacturing sector makes the greatest contribution in personnel and facilities, it needs to understand how such costs yield significant benefits. More importantly, a plant manager needs to understand such a program so that he or she, together with the QC manager and production supervisors, can respond to the results with appropriate disposition of product and changes in ingredients or process. Manufacturing and operations have the closest perspectives on product, ingredients, and process. The contribution of personnel (QC, panelists, plant coordinator) and space is critical to the success of the program.

Facilities

Because the test controls for most sensory studies are rigid in terms of the environment, as well as sample and data handling, it is necessary to provide appropriate facilities for a QC/sensory program.

Environment

In order to permit panelists to concentrate on their evaluations, the facility in which the tests are conducted should be:

- Free of noise and interruption.
- Free of odors, if aroma, flavor, or fragrance evaluations are required.
- Appropriately controlled in terms of lighting and temperature, humidity, and so on.
- Comfortable in terms of space and seating.
- Appropriately designed for the method and number of panelists.
- Located with easy access for the panelists. Proximity to the locker rooms, main entrance, cafeteria area, and/or QC laboratory generally provides easy access. (Refer to *Physical Requirement Guidelines for Sensory Evaluation Laboratories* ASTM STP 913, for guidelines for design and construction of facilities).

Some plant facilities do not lend themselves to easy design of an adequate sensory facility. Manufacturing odors may be impossible to remove from any area in the plant. The "best" available location may not meet requirements for comfort and/or proximity to panelists.

Once the current resources and the cost and potential for upgrading to a suitable facility are assessed, the quality team needs to decide if implementation at the plant(s) is tenable or if testing needs to be done at another site, possibly at R&D or nearby the plant facility.

Sample Handling

To comply with test protocols for sample handling, the test facility should be equipped with the kitchen appliances, laboratory equipment, and/or office equipment necessary to prepare and store samples for evaluation. The area designated for sample preparation must be separate from the evaluation site so that panelists cannot view the preparation process. For each test method, a strict protocol for sample handling must be written and adhered to by the QC/sensory analyst and technicians at each plant.

Outside Resources

Throughout the design and implementation of a QC/sensory program, some technical capabilities may not be available within the company, the division, or the plant at which the program is initiated. Therefore, outside expertise may be sought to insure proper development and implementation of the program.

The following technical aspects of the program, which may need outside support, are outlined with possible resources.

1. Program and methodology design can be supported by the corporate and/or R&D sensory evaluation group within a medium or large size corporation. If R&D support is unavailable, outside sensory consultants can provide QC/sensory management input for the program design and early implementation. Inside resources need to be developed within the QC/sensory staff, to interface and ultimately assume responsibility for the routine operation of the program.
2. Panelist selection and training require experienced sensory professionals, to insure that all panelists have the necessary initial acuity and develop the necessary evaluation skills. If the R&D sensory group is available to provide this expertise, one or two sensory analysts with skills in selecting and training panelists should be able to support the QC/sensory effort. Otherwise, outside consultants with panelist training skills are needed for panel selection and training.

    In this phase, outside resources (either R&D or consulting groups) are used to develop the skills of one or more sensory analysts from within the QC/sensory program. In doing this, outside resources may be needed for only one product training or one plant training by the outside sensory support. Thereafter, if the technical expertise has been transferred, further development of the program is administered directly by the plant QC personnel.

3. The test facility design should be incorporated as part of the program and methodology design. However, additional input by corporate or consulting architects and engineers is recommended, to insure the best use of space at the lowest cost to the company.

4. A system to input and analyze data can be developed and outlined as part of the total program and methodology design (No. 1 in this list). Actual implementation requires support from statistical and systems professionals, a management systems function in the company, or outside systems and/or statistical consultants. Since the QC/sensory program is unlikely to develop full systems expertise, ongoing support is likely to be required and should be an anticipated cost.

5. Correlating sensory and instrumental data requires professionals with an understanding of sensory data, instrumental data, and statistics. R&D statisticians are often capable of providing this support. If the company has no inside professionals, it is critical that the QC/sensory team seek an outside consultant who has the experience described above and who can be available for additional support, as the program develops.

In each aspect of the program that requires technical support, the sensory coordinator and plant sensory analysts must insist on support *and* training so that the QC/sensory program can become as self-sufficient and efficient as possible.

Collaborating Groups

As discussed in several instances above on page 35 on outside support, several groups within a company are needed as partners in the QC/sensory program. From each broad sector within the company, different support and collaboration are required to best utilize the company's resources.

Research and Development

- Research and development provides key technical support.
- Sensory evaluation expertise is likely to be sought from the R&D sensory evaluation group. If the group cannot dedicate one or more professionals full time to the QC/sensory program, it can provide guidance on the design of the program, the recommended outside resources or support, the identification of a sensory coordinator and plant sensory analysts, and the design of facilities and sample handling.
- Bench chemists and other product development professionals contribute infor-

mation to the QC/sensory program regarding product behavior and other technical aspects of product tolerance and product sensory attributes.

* Analytical chemists and rheologists are primary resources for recommended analytical methods that may correlate with sensory data and may be implemented ultimately at the plant level.
* Systems and statistical support are critical for developing and implementing the system to handle and analyze the data generated routinely in the QC/sensory tests.

Marketing and marketing research

Marketing and marketing research supplies information related to consumer behavior.

* Product information on sales volume, shares, consumer complaints, and marketing research test data is necessary in the early program design decision on the products to be introduced first into the program.
* Marketing research data on consumer demographics and behavior is necessary for any consumer research conducted to set product specifications.

Manufacturing

Manufacturing provides support with personnel and space.

* When plant and manufacturing management understand and commit to the QC/sensory program, by providing:
  * Personnel for plant sensory coordinator and panelists (a major cost);
  * QC time and personnel for picking up and storing samples for evaluation and references; and
  * Facility space and engineering to develop the appropriate testing site.
* Product line supervisors contribute to such a program as part of the overall corporate commitment to quality. Since plant middle management is responsible for many of the panelists used in testing and for the disposition of products that have been tested, they are on the "front lines" of the QC/sensory program. Since they routinely pay the biggest prices toward quality, they need to be informed early and throughout the program as to the benefits.
* All plant personnel are involved in the QC/sensory program, since they produce the products that are evaluated. When the objectives and results of the QC/sensory program are communicated to plant personnel, the quality message is carried to the culture of the plant.

QA/QC

QA/QC integrates the QC/sensory program into the overall QA/QC program.

* Whether the QA/QC and QC/sensory program are part of manufacturing or part of an independent quality team, the QA/QC personnel need to incorporate the

sensory program into the routine daily evaluations, the production trends, and the overall final measure of product quality.

- QC at each plant is likely to have final responsibility for interpreting the sensory results and disposing of the product as part of the regular QC function. Each plant QC manager and the product line QC supervisors must understand how the QC/sensory program works so that they use the results most effectively.

INITIAL PROGRAM STEPS

Identification of Production Variability

Once preliminary decisions on the program parameters have been made (see "Methodology", pg. 29) and the required resources and needs have been identified (see "Personnel", pg. 31), the initial program steps can be undertaken. These are:

- Identify and document the production variability of the product(s) considered for the program;
- Assess the variability data for selecting methods and approach; and
- Identify other program parameters

Importance
An in-plant sensory capability is implemented, in response to management's commitment to establishing quality and consistency in production, the lack of consistency observed in a product's sensory characteristics, and/or the possible sensory-related consumer complaints associated with that variability. It is the objective of a QC/sensory program to measure and control that variability that is not or cannot be measured by instrumental methods and, in turn, reduce the consumer complaints caused by the lack of consistent production. Consequently, the basis of the QC/sensory program is to identify and document that sensory variability.

Usually, some information on production variability is available during the planning stage of a QC/sensory program. The information that companies have on hand includes variability measurements that are documented in terms of instrumental parameters. This information is useful if the correlation between instrumental and sensory measurements has been investigated and found to exist. Otherwise, a formal evaluation and documentation of the variability in sensory terms is needed. The results of this sensory assessment will indicate 1) the sensory attributes of a product that are constant throughout production days and schedules, 2) the sensory attributes that vary over wide or narrow ranges, 3) the frequency of occurrence of that variability, and 4) preliminary information on existing fluctuations in sensory attributes dependent upon plant, production practices, shift, or ingredient and/or processing schedules.

Identifing a product's variable attributes allows the scope of a complete sensory program (such as a program operated for R&D purposes) to be narrowed to the more focused scope of an in-plant sensory capability. Compared to an R&D program, which measures each sensory attribute of a product in detail, the in-plant program will focus on only those attributes that present extreme fluctuations during production and/or whose variability affects consumer acceptance.

Procedure for Variability Assessment
The evaluation of a product's variability is completed through the combined effort of the established in-plant QC department and sensory professionals.

Sampling of Products
The quality control professionals assist in collecting the samples required for the evaluation. They supervise the product's sampling plan to assure that all ranges of processing conditions are represented during the collection of products to be surveyed. These might include different raw materials suppliers and/or batches, processing equipment, processing parameters, and environmental conditions. The available instrumental data base on production variability and current sampling plans might be sources of information used to determine the sampling plan required for the collection of production samples.

To study the variability of one or more products or brands across one or more plants, QC professionals and the sensory coordinator need to decide on the scope of the sampling. From earlier data, the QC department should have a sense of the period of time (one day, one week, one month) across which the product tends to manifest a normal range of variability. Once this time period is identified, samples for the study are pulled from the broadest array of sources of variability.

These include the following:

- Set time intervals *within* the time period to be studied:
  - Every hour across a day; or
  - Every day across a week.
- Designated plants that produce the product:
  - All plants;
  - Two plants with batch processes or two with continuous; or
  - One plant that produces the bulk of the product.
- Specific shifts within a day:
  - All shifts; or
  - One key shift that has demonstrated high variability.
- Pickups within a shift:

- Early and late in a shift;
- At key points in a shift with batch changes; or
- Once midpoint in a shift.

Decisions regarding the number and scope of these pickups determine the potential for identifying the range and frequency of variability in a product's production.

There are no set criteria as to the exact number of samples and the sampling plan recommended for the collection of products, since the production variability varies from product to product and manufacturer to manufacturer. Without considering spot problems or unusual processing practices, it has been the authors' experience that, on the average, the collection of products from three weeks production usually provides an adequate and representative picture of the usual production variability. Such sampling criteria are valid for products such as foods (baked goods, salad dressings, ice cream, or confections) or personal care products (mouthwash, deodorant, or shampoo) that are produced in a few hours or less. For products requiring days or weeks to produce (wine, oven dried fruits, pickled vegetables), the sampling covers a wider time period than a few weeks. It is advisable that the survey be conducted on two consecutive weeks and that the third week be completed on a month following the initial survey. This survey might be reduced slightly when the survey includes sampling production from more than one shift. Following this or other similarly effective sampling practices, approximately 150 to 250 products should be available for sensory screening.

Sensory Assessments of Products

The following plan is suggested as one way to determine product variability in terms of the sensory attributes that vary and the size of the variability for those attributes. The evaluation of the production samples collected for the identification of the product's production variability requires the expertise of a trained descriptive panel. This support is usually provided by the R&D sensory evaluation department, if a descriptive capability is available. Otherwise, this service is contracted.

Initially, the evaluation process requires the full descriptive characterization of two or three of the production samples by a descriptive panel (10 to 15 people). A full descriptive analysis of these samples represent the evaluation of 30 to 40 sensory (e.g., appearance, aroma/fragrance, flavor, texture, skinfeel, or tactile) attributes that fully characterize the specific product and indicate the level/intensity of each of those attributes. Figure 2.1 shows the Spectrum™ Descriptive Analysis characterization of a particular peanut butter production sample. This characterization represents the benchmark for the ensuing evaluations, where only a subgroup of the panel (3 to 7 people) is required. This panel uses the full

```
APPEARANCE
Color intensity    |_____x_____|
Color chroma       |_____x_____|
Surface gloss      |_____x_____|
Surface graininess |____x_____|
Oil separation     |__x_____|
FLAVOR
Roasted peanutty   |_____x_____|
Raw beany          |_____x_____|
Dark roast         |_____x_____|
Sweet aromatics    |_____x_____|
Woody/hulls/skins  |_____x_____|
Cardboard          |__x_____|
Painty             x_____|
Fruity fermented   |____x_____._____|

Sweet              |_____x_____|
Salty              |_____x_____|
Bitter             |_____x_____|
Sour               |____x_____|

Astringent         |_____x_____|
TEXTURE
SURFACE
Graininess         |__x_____|
Oiliness           |____x_____|
Stickiness         |_____x_____|
FIRST COMPRESSION
Firmness           |_____x_____|
Cohesiveness       |_____x_____|
Denseness          |_____x____|
Graininess         |__x_____|
Stickiness         |_____x____|
BREAKDOWN
Mixes with saliva  |_____x_____|
Cohesiveness of mass|_____x_____|
Moistness of mass  |_____x_____|
Graininess of mass |__x_____|
Uniformity of mass |_____x_____|
RESIDUAL
Chalky film        |____x_____|
Grainy particles   |__x_____|
Oily film          |__x_____|
x = product mean;  range of production falls between the vertical slashes
*  Products are rated on a 15 cm. line scale based on the
       Spectrum Descriptive Analysis Method
```

FIGURE 2.1.   Peanut butter category survey. Products are rated on a 15-cm. line scale, based on the Spectrum Description Analysis Method. x = product mean.

descriptive Spectrum™ (Fig. 2.1) to proceed with the screening of the remaining 150 to 250 samples over a series of sessions and days. Trained panelists evaluate these products (e.g., taste, feel, smell) by focusing on the attributes that differ in intensity from the benchmark evaluation (Fig. 2.1).

Documentation of Variability

Figure 2.2 shows the final results of this screening process: 1) the list of the variable sensory characteristics (representing a subset of all attributes that characterize that product) and 2) the corresponding variability (intensity) ranges. The evaluations differentiate those characteristics presenting wide variability ranges (e.g., saltiness) (Fig. 2.2) from those with a narrow variability band. (e.g., stickiness). In addition, intensity histograms for each of the variable attributes are developed (Fig. 2.3). These histograms summarize the percentage of samples for each intensity in the variability range. This information is used in selecting a control, as explained below.

Depending on the product type being evaluated, this process might take several days to complete. The need for valid and reliable data on the identification of variable attributes dictate that those evaluations be conducted under the most strict test controls possible, including a close attention to sample presentation conditions, number of samples evaluated per session, rinsing materials, and so on (Meilgaard, Civille, and Carr 1987).

Uses of Variability Information
The variability information collected at these initial steps determines many program characteristics. Among these are:

- The need for a QC/sensory program. If a wide production variability exists, a recommendation is made to establish and operate a program as initially defined. If production fluctuates only slightly, the initial plans for the program implementation are modified either to develop a less involved program or to continue operating with the instrumental techniques already established.
- The attributes that are measured and controlled. Consumer acceptance data are collected and show those attributes that are both variable and have an impact on consumer acceptance.
- The type of method selected. Depending on the type of attributes that vary, their degree of variability and other considerations discussed before (such as the type of measurement that is most compatible with the company's philosophy), the final sensory method is selected.
- The type of sensory specifications needed. The results on production variability have an impact not only on the type of sensory method selected, but also on the type of specifications needed.

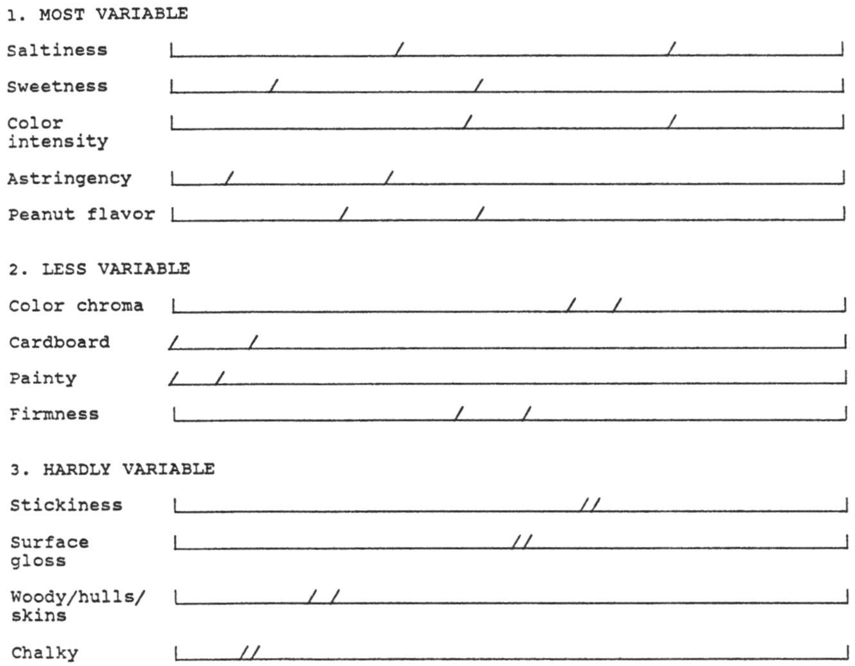

1. MOST VARIABLE

Saltiness

Sweetness

Color intensity

Astringency

Peanut flavor

2. LESS VARIABLE

Color chroma

Cardboard

Painty

Firmness

3. HARDLY VARIABLE

Stickiness

Surface gloss

Woody/hulls/ skins

Chalky

FIGURE 2.2.   Attributes with different ranges of variability.

- The type of consumer research needed to set specifications.

The next step is the consumer research study, whose objective is to determine the effect of these variable attributes on consumer acceptance. The consumer test is designed to collect acceptance data, as well as diagnostic information on the variable attributes (e.g., liking and consumer-perceived intensity of peanut flavor). This consumer information is used to set the sensory specifications used for decisions for the daily disposition of regular production.

The procedures to design and conduct the required consumer test and to analyze the data and establish sensory specifications is covered in detail in Chapters 3 to 6, for the four sensory methods recommended.

Selection of Test Methodology

Based on the type and size of variability in production and the capacity to ascertain consumer acceptance related to that variability, the QC/sensory professional deter-

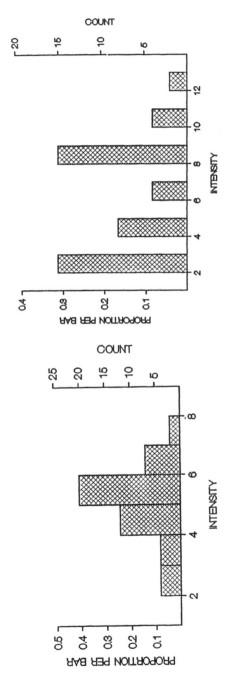

FIGURE 2.3.   Histograms for incidence of intensities/levels of one key attribute.

mines the most effective sensory method to use as part of the ongoing day-to-day evaluation of samples.

The decision regarding which method is best requires some judgement based on the following criteria:

- If the variable attributes can be or have been determined and if these are limited to five to ten key attributes, the comprehensive method is feasible. Yet it requires careful training, using good references and dedicated panelists.
- In cases where a specific product control can be identified and where product variability is limited to a single continuum, such as amount of processing, but where this variability is not easy to characterize with specific attributes, the difference-from-control method is a likely method of choice. Complex products, such as coffee, beer, yogurt, and so on, which involve raw material compounded by processing variables may fall into this category.
- When the product variability is not easily defined by specific sensory attributes, but can be more readily reflected in the broad sensory parameters (appearance, flavor or fragrance, texture), the quality rating method is a likely method of choice.
- In those cases, when variation cannot be specifically defined by attribute, when no single control is identified, and when examples of unacceptable product cover a multitude of sensory conditions, the in/out method is recommended.

There are, however, cases in which the QC/sensory professional has more than one option or in which the choice of test method may not be clear cut:

- In the absence of the resources or expertise to conduct consumer research, management (preferably manufacturing, marketing, R&D, and QC) can replace the consumer acceptability based specifications with a set of management specifications that define the acceptable range of each attribute. If specific examples of the ranges across the variable attributes are not available for management evaluations, use of the comprehensive method may be thwarted. If, on the other hand, some examples of product with appearance, flavor, or texture that are not "typical" or "acceptable" by management standards can be collected, these may become the basis for using quality scaling. In such cases, references or examples can more easily be created by R&D doctoring or stressing product to develop broad "off" or odd appearance, flavor, or texture, which can be used in the simpler training for quality scaling. Only two or three samples for each broad sensory category are generally necessary for setting specifications based on quality scales.
- Not every product has a complex spectrum of attributes that can be readily described. Consumer complaints may consistently reflect a single sensory defect. The variability may be identified in one meaningful dimension, such as an

off-odor in an unscented liquid detergent. The difference-from-control or in/out methods are sufficient to assess this type of variability.

If it is possible to obtain and keep under controlled conditions 1) a "control" that exhibits no or low level off notes and 2) some examples of the off-note characteristic at different intensities, it is possible to train a panel to know the control and an array of samples different from the control. With these references, or anchors, the panel can establish a degree of difference scale from the control and use it to evaluate regular production. Assessing consumer acceptance relative to these different degrees or levels of difference from the control allows the QC/sensory professional to set a limit on the size of difference that is acceptable. If conditions are such that only a few examples of production with the off-note are shown to a panel vis-a-vis a "clean" sample (one with little or no off-note), the in/out method may be preferred. The lack of range of difference precludes the difference from control method and suggests the in/out alternative.

It is unlikely that any QC/sensory program will find it best to use *only one* sensory test method for all products and raw materials.

Identification of Other Program Parameters

Identification of a Control
For the implementation of a QC/sensory program, where judgements are made relative to a "control" or a "standard," identification and documentation of such product is required during the preliminary phases of the program. This step is usually completed in the evaluation process for identifying production variability, described earlier.

A "control" used for quality purposes is referred to as a product that is used as a representation of certain characteristics (not necessarily the "optimal") and a product that can easily be obtained, maintained, or reproduced. The criteria for choosing a "control" can be arbitrary or deliberate. For example, a "control" might be selected as 1) a product manufactured in plant X, 2) the product with the characteristics produced "most frequently" (the product in Fig. 2.3 a), 3) the product preferred by consumers, 4) the product with a given intensity of an attribute, or even, 5) the pilot plant product.

A "standard" is a product considered to be the "optimum," "most preferred by consumers," or the highest quality product a company can manufacture. In some companies, this is called the "gold standard." The gold standard has value in a company versus an ultimate target for R&D and manufacturing to work toward producing. Since it does not represent current normal production and since it may not be possible to produce it using current materials and processes, the gold standard should not be used as the control in a QC/sensory program.

It is clear with this distinction that the selection of a "standard" is more involved and complex than the selection of a "control." A considerable amount of research and the input of management are required to select and document a "standard" product. As a result, few consumer products companies have identified a "standard" for judging the overall sensory quality of daily production. In many cases, companies find that using such standards is frustrating to manufacturing and R&D, since they can be reproduced only under the very best, and thus limited, circumstances. This discussion shall only focus on selecting and using a "control" for quality control purposes.

Two philosophies exist among companies regarding *who* determines what is "in spec" or "good." Most companies realize that it is the consumer who dictates which products or product characteristics should be considered "good quality." Others rely on upper management's input to set the criteria. With some products, consumers have rather loose criteria for "goodness" and find the bulk of a company's production acceptable. Management may choose to set stricter criteria as a commitment to quality, in spite of the consumer's broad acceptance (i.e., quality = consistency).

Independent of the criteria selected to establish a "control," a survey, as described in the "Identification of Production Variablity" section, on page 40, is needed to provide supportive information when selecting a "control." Specifically, the information on product variability (Fig. 2.2 and 2.3) is needed to select the product that is considered a "control." Management needs to know the position of various potential "controls" in the production variability ranges and their frequency of occurrence, to select a given product as a "control." For example, it is unlikely that a company will select a product with an intensity of 6 in attribute 2 (Fig. 2.3) as a "control," even though it would be found to be the product most liked by consumers. The frequency distribution graph shows that a product with this intensity is produced very infrequently and would not represent an adequate control. If this product is chosen as a "control," a large percentage of daily production would be rejected, since product close to the "control" is produced very infrequently. A company would most likely select the next highest quality product as the "control" (e.g., a product with intensity of 8).

Additional Uses of Variability Data

In many situations, the sensory information provided to management and operations, regarding production variability, is used to assess the production process prior to implementing a sensory program. This assessment relates the results of the sensory surveys to specific causes, such as origin (production facility), operating conditions, ingredient sources, and so forth. Developing an understanding of sources of variability is recommended, particularly for infrequently occurring problems. This assessment can identify uncommon erratic sources of materials or production practices that can be eliminated prior to implementing the sensory program.

Identification of Raw Material Variability

The results of a study of finished product variability may indicate the need to concentrate on the variability of raw ingredients. This leads to the development of a program specifically designed to control a product from the raw material stage. Control of raw materials as a first step to quality control is highly recommended, especially if those raw materials have distinct and variable characteristics of their own.

Raw materials, such as fragrances, herbs and spices, and commodities that develop flavor with processing (peanuts, cocoa and coffee beans, potatoes, etc.) can be treated much like the finished products discussed in this chapter. The QC/sensory program can ascertain variability over a time period and set specifications. The specifications can be based on the raw ingredients' sensory properties tested at the time they are received at the plant or based on the sensory properties they eventually produce in finished product.

Special experiments are needed when the raw materials are added to the process before key sensory properties develop. Studies must relate properties of the raw material or in-process product to the sensory characteristics of the finished product. Then, specifications on the raw materials can be set, to insure that the ingredients or raw materials are not driving the sensory attributes of the finished product "out of spec."

Such experiments may initially seem costly and complex. They are, however, highly recommended as a means to controlling quality *and* costs. A poor raw material, rejected and sent back to the supplier, costs only the time for its evaluation. However, a poor raw material, incorporated into a product, may result in a large rejected segment of production that involves a much greater cost.

Summary

The decision to implement a QC/sensory program requires a considerable amount of "up front" work before the program is established. The question of where to start requires study of

1. Products, in terms of sensory problems and variability;
2. Process, in terms of raw materials, in-process, and finished products;
3. The attributes to study, in terms of variability and effect on consumer acceptance; and
4. Sampling, in terms of the rate and amount.
   Identification of program parameters or needs must be balanced with identifi-

cation of the resources required to meet those needs. Time, money, and corporate support are required to provide the test methods, personnel, test facilities, and support services required to implement the program. As with all sensory testing, the objective is to provide the most sensory information in a timely and cost-effective way. Balancing needs against resources early in the program is essential.

The first steps in the program are to determine the product variability in terms of type, frequency, and size and to select the appropriate test method(s), to effectively and efficiently track that product variability. Cooperation between QC and sensory personnel is critical during this phase, since each group brings different perspectives to the identification of program focus and program resources.

3

Comprehensive Descriptive Method

ABSTRACT

This approach is the most comprehensive in-plant sensory program. It represents a program in which a well-trained sensory panel operates as any other analytical instrument in the QC laboratory. This panel provides data on the intensity/level of a small set of the product's sensory attributes (approximately 5 to 15). These attributes are known to vary during production and have been found to affect consumer acceptance. A specification is set for each of the variable sensory characteristics.

Specifications represent the tolerable range of intensities for plant production. Products whose intensity on any given attribute fall outside those set specifications are considered unacceptable. The specifications are set preferably with the input from consumers and management and the consideration of realistic production and cost limitations. Due to its comprehensive and complex nature, this program is mainly geared to the evaluation and quality control of finished products.

The two main advantages of this approach are 1) the absence of any subjectivity in the evaluation (panelists act as sensory instruments, without incorporating non-product information that might bias their judgements), and 2) the quality of the data obtained. These data lend themselves to a variety of data analyses and manipulations, such as those that show the relationship of the panel data (product's sensory data) with either instrumental and/or consumer data. This program's main disadvantage is the high cost and the time involved in its implementation and operation.

PROGRAM IN OPERATION

The comprehensive descriptive approach consists of having at each of the manufacturing facilities a well-trained sensory panel that evaluates daily production samples. All panels are trained similarly and provide comparable information for any given sample. Each panel provides information on the intensity/level of the critical sensory variable attributes of the product.

Table 3-1 shows the results of the evaluation of a production batch of potato chips labelled 825, using this method. The results shown are the average value of the panel scores for each attribute (e.g., 7.5 for hardness) and are interpreted as any other analytical/instrumental information would be. They indicate the level at which each of the product's attributes is perceived.

These data, provided to QC management, are used to make decisions regarding the disposition of the evaluated production batch. These decisions are based on the comparison made between the panel results and the specifications set for each of the product attributes. The column on the right-hand side of Table 3-2 shows the sensory specifications. The sensory specifications for each variable attribute were added to this table for the purpose of this discussion. However, it should be clear that those specifications are never included in the panelists' evaluation forms (ballots) and should not be information provided to panelists. This information is known only by management, to compare the panel results to the product's specifications. A product is considered unacceptable if it falls outside the specifications

Table 3-1.   Average panel results for "batch 825" of variable potato chip attributes

Attribute	"Batch 825" Score
APPEARANCE	
Color intensity	4.7
Even color	4.8
Even size	4.1
FLAVOR	
Fried potato	3.6
Cardboard	5.0
Painty	0.0
Salty	12.3
TEXTURE	
Hardness	7.5
Crisp/crunch	13.1
Denseness	7.4

Table 3-2.   Comparison of panel results with sensory specifications

	Panel Data (sample 825)	Sensory Specifications
APPEARANCE		
Color intensity	4.7	3.5–6.0
Even color	4.8	6.0–12.0
Even size	4.1	4.0–8.5
FLAVOR		
Fried potato	3.6	3.0–5.0
Cardboard	5.0	0.0–1.5
Painty	0.0	0.0–1.0
Salty	12.3	8.0–12.5
TEXTURE		
Hardness	7.5	6.0–9.5
Crisp/crunch	13.1	10.0–15.0
Denseness	7.4	7.0–10.0

("out-of-spec"). For example, Table 3-2 shows that the production sample 825 would be considered unacceptable or out-of-specifications. The intensities of evenness of color (4.8) and cardboard (5.0) fall outside the tolerable intensity ranges set (specifications: 6.0 to 12.0 for evenness of color and 0.0 to 1.5 for cardboard). The analysis of these data help management in their decision-making process.

If an SPC program is being used (see Appendix 5), the data are plotted on control charts and determination of "in-control" production is made. The trend toward an out-of-spec cardboard intensity, for example, may have been identified on the control chart before product had to be discarded (see Fig. 3.1).

IMPLEMENTATION OF THE PROGRAM

A company that selects the comprehensive descriptive method for a QC/sensory program has the following objective and has collected the following information in preliminary stages (Chapter 2).

1. Overall objective: To establish a program for evaluating the quality of finished products (and/or complex raw materials) that uses a highly trained panel to measure the perceived intensity/level of selected product attributes. These results are compared to a set of specifications for each attribute.
2. Preliminary information and special considerations:

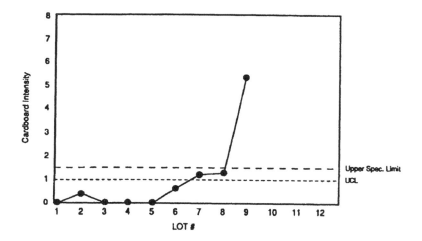

FIGURE 3.1.   Control chart of cardboard flavor intensity of potato chips, showing the increasing trend before the out-of-paper spec batch 9 was produced.

a. Production samples have consistently exhibited perceivable differences (i.e., variability) over time.
b. The differences have been identified in terms of the sensory attributes and the variability (intensity) ranges of those attributes.
c. Products representative of these production variations can be easily obtained in any product survey, or produced otherwise for training purposes.
d. The resources to implement this type of program are available at both the production and research facilities (refer to Chapter 2 "Identifying Resources," pg. 28).

The four steps in implementing this type of program are:

1. Establish sensory specifications;
2. Train and operate of the plant panel(s);
3. Establish a data collection, analysis, and report system; and
4. Operate the ongoing program.

Establishment of Sensory Specifications

As shown in Table 3-2, the sensory specifications for the comprehensive approach are represented by the range of intensities accepted or tolerated for each "critical" attribute—that is, "in spec." production falls within the intensity limits set by the sensory specifications. Conversely, products falling outside the specifications are considered unacceptable. For example, the sensory specification set for crispness/crunchiness in potato chips (Table 3-2) indicates that production samples with crispness intensities lower than 10.0 are to be considered unacceptable or out of specification, with an ensuing rework or rejection decision.

There are two ways by which sensory specifications for the comprehensive program are established. The preferred approach is for the specifications to be set through consumer and management input. Alternatively, specifications can be set based on management's criteria alone. A description of both procedures follows.

Specifications Set Through the Input of Consumers and Management

Consumer acceptance is driven by certain, but not all, product characteristics (Barnes et al. 1991). These characteristics vary from product type to product type and sometimes from brand to brand. Companies often lack such consumer data. More typically, the information available is the opinion of upper management or the brand group as to what factors are critical to a product's acceptance. This information can deviate from the actual consumer's response to a product.

Product testing with consumers is required to obtain direct information on the attributes that drive a product's acceptance and on the consumers' tolerance to variability of such attributes. This research involves selecting and testing a series of production samples that represent the extreme and intermediate points of the attributes' ranges. These data are used to establish more realistic sensory specifications.

A product test for establishing sensory specifications involves the following seven steps:

1. Assessment of needs and planning phase;
2. Collection and initial screening of a broad array of production samples;
3. Descriptive evaluation to characterize the samples;
4. Selection of samples to be tested by consumers;
5. Planning and execution of consumer research test;
6. Data analysis to establish the relationship between the consumer and descriptive data sets; and
7. Establishment of the final specifications based on test results and management's input.

Assessment of Needs and Planning Phases for the Cosumer Research Study

Resources. The approach presented in this section and followed by the authors requires the following resources:

- A sensory coordinator responsible for the planning, administration, and completion of all steps;
- A sensory professional (R&D or consultant) to assist the sensory coordinator in the implementation stage;
- Professional support from R&D and the plant to provide support in consumer testing, data analysis, and so forth;
- A descriptive panel;
- Computer resources; and
- The funds to plan, administer, and execute the whole project.

Product Amount and Storage.   The number and type of sensory tests determine the total amount of product needed. Although this is important in all sensory test designs, it is critical in a quality control consumer research situation because of the large amount of product handled and the amount of product storage needed until all testing is completed. The large volume of product is a function of all the tests to be conducted (product inspection for screening, descriptive, and consumer tests) and all the products that are initially collected and surveyed.

Table 3-3 shows an example of the amount of product needed to establish specifications for the potato chip example. Part I of the table corresponds to the common calculation usually followed when planning sensory tests. Part II shows the unique characteristics of the calculation of product needed for quality control purposes. For a comprehensive descriptive program, all the amount in Part I of Table 3-3 must be multiplied by the total number of production samples collected. If the recommendations of Chapter 2 are followed and 100 to 150 product samples are collected over one to four weeks of production, then the total amount of product to be collected and stored is 100 to 150 times the amount calculated in Part I of Table 3-3. For the potato chip example, the total amount of product with 100 production samples is 8,500 bags of chips.

The large storage space is needed for a short time only. Once the final sample selection for the consumer tests has been made, up to 70% of all the product can be either placed back in the distribution cycle or disposed.

Test Scheduling.   The most important factor to consider when scheduling all tests (e.g., product screening based on variability, consumer, and descriptive tests) is the shelf life of the product being studied. When shelf life is not a limitation, all tests can be planned and conducted leisurely. Among these cases are the testing of paper products, fabrics, and some household and personal care products. However, for other products—certain food and beverage products are extreme cases of this category—a careful scheduling of activities is required, due to the product's short shelf life. The schedule has to assure that the products to be tested undergo minimal, if any, changes from the time they are produced or distributed until they are evaluated. This problem is alleviated in those few cases where pilot plant

Table 3-3.   Calculation of product needed for setting specifications

Potato Chips

I. Amount of product needed for each production sample collected

		Bags 7 oz.
• Identification of production variability		
1st inspection		1
2nd inspection (for selected samples)		1
• Descriptive analysis (10 panelists)		
(initial) 1st rep		2
2nd rep		2
(2nd evaluation) Close to consumer study (optional)		(2)
• Management meeting		
20 participants		2
• Consumer study		
100 consumers per location	(20)	
300 consumers per location		60
A. Minimum units per sample		68
B. Total required (A + 0.25A)		85

II. Total amount of product to be collected and stored

• Units per pickup	85 bags
• Total units collected for 100 pickups	8,500 bags

produced samples are similar to production samples. Then, pilot plant samples can be specifically produced for the consumer test, without any scheduling problems.

For products with relatively short shelf lives (e.g., up to six weeks) and no pilot plant substitutes, the scheduling of all evaluation tests is difficult, if not impossible. In this case, all the product evaluation tests are to be conducted in two phases. Specifically, the first phase includes only the sample collection, screening, and documentation of variability. The second phase consists of screening samples and the consumer and descriptive tests.

Sample Collection and Initial Screening

The type of sample collection needed depends on the history of the QC/sensory program, specifically the amount of surveys completed in the past. If no survey and identification of variability have been completed (Chapter 2, "Identification of Production Variability", pg. 40), a broad sampling needs to be planned and administered. On the other hand, if other surveys have been completed, the sampling needed is less involved and is geared to find the samples that span the entire range of production variability.

   Complete Survey.   The following survey is scheduled for the potato chip example. Product is collected from:

- Three manufacturing sites;
- Nine production days;
- Two shifts; and
- Two pickups within a shift.

This survey leads to a total of 108 production samples (pickups) to be screened. Through the screening process, a subset of these 108 samples is selected for further testing. Samples that do not show extreme variability are eliminated. Samples that show variability from what is known be "typical," or from the control, or from the majority of the production samples being evaluated are chosen. In general, no more than 50% of the total number of products is selected through this initial screening. For the potato chip example, a total of 25 products are selected.

   Smaller Scale Survey.   In those cases where the product variability has been identified through previous surveys, a smaller scale survey is completed. The objective of this survey is to collect enough samples that display the variability of all sensory characteristics.

   In the small scale survey, fewer pickups are needed. For example, instead of 108 pickups for the potato chip project, only 60 or 80 pickups are necessary. In addition, based on previous information, additional pickups from specific plants or shifts can be included, if they are known to show large variability in previous surveys.

   As in the complete survey, 25 products that represent the extremes and span the entire range of production variability are selected for the potato chip example.

Descriptive Analysis
The descriptive analysis is completed to obtain a detailed characterization of the selected subset of products. This information shows each sensory characteristic of the product and the intensity/level at which it is perceived. The descriptive information is used for three purposes:

1. The selection of the final set of products to be consumer tested (pg. 60);
2. The establishment of specifications (pg. 76); and
3. The selection of replacement product references in later stages of the program operation.

The descriptive information is obtained from a trained panel. This test can be completed through the R&D sensory panel, if there is one in the company. Otherwise, a contract research panel can conduct these evaluations.

An experienced sensory professional designs and administers the descriptive test. Consideration should be given to the experimental design, sample preparation and evaluation procedures, and test controls (Meilgaard, Civille, and Carr 1987).

A complete characterization of each sample includes the evaluation of a total of 20 to 40 attributes per sample. Table 3-4 shows the list of all attributes evaluated for potato chips and the results for one of the batches evaluated. (The scale is a 0-15 intensity scale where 0 = none and 15 = extreme).

Upon completing this test, a similar descriptive characterization is obtained for each of the 25 potato chip samples screened. With this test, more precise information on the product variability is obtained. Table 3-5 shows the summary statistics and its variability ranges for each attribute obtained through the evaluation of 25 samples. An inspection of these results gives an indication of which attributes present small, medium, and large variability. For example, some of the attributes showing the largest variability are evenness of color, evenness of size and shape, saltiness, and crispness.

The descriptive characterization is used for the final sample selection described below and for the data analysis set for setting specifications, described later in this chapter (pg. 76).

The same procedures and principles that document production variability, as described for the potato chip example, would apply to this or other food and nonfood consumer product. For example, Figure 3.2 shows the production variability ranges for nail enamels. The line scales are 15 cm scales, where the distance between the left end marked "0" and the first slash mark on the line represents the lower range intensity value. For this product, the attributes that show high variability (e.g., opacity, spread) can be identified from these results.

Sample Selection

Starting with a collection of samples that span the range of typical product variability, it may not be clear as to which attributes best summarize the variability, nor which subset of samples most economically span the variability in all of its meaningful dimensions. It turns out that the two issues are intrinsically linked. Identifying attributes that exhibit meaningful variability and selecting representative samples occurs simultaneously.

There is no fixed analytical procedure to accomplish the task of identifying variable attributes and representative samples. The approach is instead an exploratory process in which samples possessing unique (and not so unique) combinations of attribute ratings are identified while simultaneously tracking the varying and co-varying of the attribute ratings themselves. Sophisticated data analysis

Table 3-4.   **Descriptive characterization of a potato chip production sample (613)**

Attribute	Sample 613 Score
APPEARANCE	
Color intensity	3.2
Evenness of color	6.0
Blotches	0.0
Translucency	9.7
Evenness of size	3.4
Evenness of shape	3.6
Thickness	4.5
Bubbles	5.2
Folds	7.3
FLAVOR	
Potato	5.3
Raw	2.1
Cooked	2.0
Fried	3.3
Skins	0.7
Heated Oil	1.5
Earthy	0.8
Cardboard	1.2
Painty	0.0
Salty	10.5
Sweet	4.3
Bitter	0.5
Astringent	5.0
Burn	4.0
TEXTURE	
Surface bumpiness	5.2
Oily surface	4.1
Hardness	6.3
Crispiness/crunchiness	12.4
Denseness	8.9
Number of particles	11.5
Abrasiveness of particles	4.5
Persistence of crisp/crunch	6.1
Mixes with saliva	10.1
Cohesiveness of mass	4.1
Grainy mass	6.7
Toothpack	6.5
Oily film	4.3

Table 3-5.   Summary statistics of the sensory attributes for 25 samples of potato chips

Attribute	Mean	Std. Dev.	Minimum	Maximum	Range
Color int.	3.2	0.5	2.0	8.0	6.0
Even color	6.0	0.9	4.0	13.3	9.3
Blotches	0.0	0.0	0.0	0.0	0.0
Translucency	9.7	0.6	8.6	12.2	3.6
Even size	3.4	1.3	4.0	8.4	4.4
Even shape	3.6	0.8	1.0	6.1	5.1
Thickness	4.5	0.1	4.4	4.6	0.2
Bubbles	5.2	0.2	5.1	5.4	0.3
Folds	7.3	0.8	5.0	10.1	5.1
Potato complex	5.3	0.7	2.6	6.7	4.1
Raw	2.1	0.2	0.7	4.2	3.5
Cooked	2.0	0.0	2.0	2.0	0.0
Fried	3.3	0.5	2.0	5.0	3.0
Skins	0.7	0.1	0.6	0.8	0.2
Heated oil	1.5	0.2	1.2	1.6	0.4
Earthy	0.8	0.1	0.7	0.8	0.1
Cardboard	1.2	0.3	0.0	5.1	5.1
Painty	0.0	0.2	0.0	3.8	3.8
Salty	10.5	1.1	8.1	14.7	6.6
Sweet	4.3	0.1	4.2	4.3	0.1
Bitter	0.5	0.0	0.5	0.5	0.0
Astringent	5.0	0.0	5.0	5.0	0.0
Burn	1.3	0.5	0.0	3.2	3.2
Surface bumps	5.2	0.1	5.0	5.3	0.3
Oily surface	4.1	0.5	2.8	5.1	2.3
Hardness	7.3	0.7	4.7	9.4	4.7
Crisp/crunch	12.4	0.7	9.5	14.6	5.1
Denseness	8.9	0.6	6.1	10.0	3.9
Number of Particles	11.5	0.2	11.2	11.6	0.4
Abras. of part.	4.5	0.1	4.2	4.6	0.4
Persist. of crisp	6.1	0.1	6.0	6.2	0.2
Mixes w/saliva	10.1	0.1	9.9	10.2	0.3
Cohesiveness	4.1	0.3	3.8	4.3	0.5
Grainy mass	6.5	0.0	6.5	6.5	0.0
Toothpack	3.5	0.5	1.5	4.7	3.2
Oily film	4.3	0.1	4.3	4.4	0.1

procedures are available to apply to the problem. However, only the minimum level of computational complexity necessary to accomplish the task should be used. Blind reliance on sophisticated techniques can yield misleading results. Simple graphical techniques play a central role in the approach and may be all that are required. The basic data analysis tools used for sample selection are data plots (histograms, scatterplots, etc.), summary statistics (means, standard deviations, correlation, etc.), and, if necessary, principal components analysis.

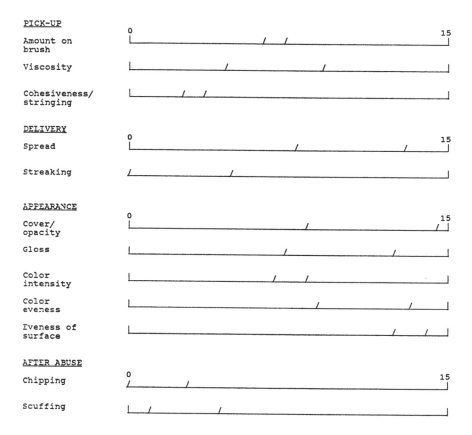

FIGURE 3.2.   Production variability of sensory attributes of nail enamels.

The approach is illustrated using the descriptive data on the 25 samples of potato chips that have been evaluated on 36 attributes by a trained descriptive panel (see Table 3-5). What follows is a step-by-step sequence of analyses used to identify variable attributes and to select representative samples for further study.

STEP 1: Screen the Attributes for Overall Variability.   The analysis begins by determining which attributes exhibit meaningfully large variability. The range of responses that occurs in each attribute is obtained by examining the summary statistics for the samples, such as in Table 3-5. Attributes that exhibit trivially small amounts of variability should be dropped from further analyses. What constitutes "trivially small" must be decided on a case-by-case basis. This is especially true for off-notes, which are likely to have substantial effects on acceptability when present at even very low levels. In the example, a total range of values of 0.5 units or less has been deemed to be too small to be worried about. Of the original 36 attributes, 18 show little or no variability. The attributes remaining after the initial screen are presented in Figure 3.3.

The absence of statistically significant differences among the samples should not be used to decide what constitutes a trivially small range of responses. Failure to find significant differences does not mean that the samples are the same. Rather large differences (e.g., 1 or 2 units on a 15-unit scale) may fail to be declared significant for a variety of reasons, including small panel size or confusion about the attribute scale. Familiarity with the product category and good sensory judgment should be used *with* the statistical results to make this decision.

APPEARANCE	0	15
Color int.		
Even color		
Translucency		
Even size		
Even shape		
Folds		
FLAVOR	0	15
Potato		
. fried		
. raw		
Cardboard		
Painty		
Salty		
Burn		
TEXTURE	0	15
Oily surface		
Hardness		
Crisp/crunch		
Denseness		
Toothpack		

FIGURE 3.3.   Production variability of sensory attributes of potato chips.

STEP 2: Screen the Samples for Extreme Ratings.   Using only the attributes that remain after step 1, attention now focuses on the samples. Histograms of the attribute ratings, such as those in Figure 3.4, are used to identify samples at the extremes of the distributions. The graphical approach will also reveal more about the general distribution of the ratings than is obtained from the summary statistics tabulated in step 1. For example, Figure 3.4a shows a regular, bell-shaped distribution of ratings for even shape. No sample stands out as odd in the plot. Similarly, in Figure 3.4b, no sample is outstanding for even size. However, the even size histogram reveals a uniform (i.e., flat) distribution of ratings. Even though even shape has a slightly larger range of variability (see Table 3-5), sample-to-sample variability in even size is more apparent because of the greater likelihood that an individual sample will fall at the extremes of the range.

Figures 3.4c and 3.4d illustrate two important situations involving "extreme" samples. In Figure 3.4c, sample 4 has a very high saltiness rating and sample 7 has a very low saltiness rating, relative to the rest of the samples. The remaining samples exhibit a regular pattern, similar to Figure 3.4a. Figure 3.4d shows another important case, where, for the attribute painty, sample 12 exhibits a moderate response (3.8), while the remaining samples in the group all have ratings of 0. The tabular summary of the data done in step 1 indicates a meaningfully large range of values for painty, but all of the apparent variability is due to a single sample.

Based on Figures 3.4c and 3.4d, samples 4, 7, and 12 will be included in the consumer test, to determine the effects of their extreme behavior. However, they will not be included in any further steps of the current sample selection analyses because doing so would reduce the sensitivity for identifying variable attributes and representative samples.

In addition to looking for extreme samples, the distributions of the attribute ratings should also be examined for any other atypical patterns, such as multi-modality (i.e., multiple 'peaks') or, in the extreme, groups of samples with completely nonoverlapping ranges of ratings. This information should be noted. Results of parametric analyses, such as correlations and principal components, must be interpreted in light of any peculiar patterns that are known to exist. Further, the information can be used directly in the final selection of samples for the consumer test.

Examining the histograms of the remaining attributes (not shown) reveals that raw potato flavor and cardboard exhibit the same behavior as painty, with all but one of the samples (either 4, 7, or 12) having the same, or nearly the same ratings. Since samples 4, 7, and 12 are already selected for consumer testing, and there is no additional meaningful variability in these three attributes, they can be eliminated from further consideration in the sample selection process.

STEP 3: Identify Samples that Depart from Interrelationships Among the Attributes.   Eliminating extreme samples and non-varying attributes from the analysis permits a more sensitive study of the interrelationships among the remain-

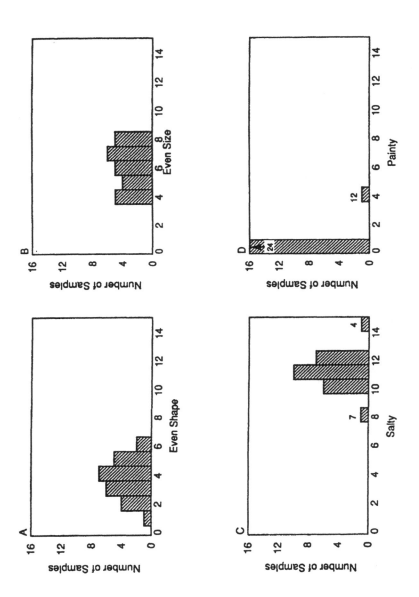

FIGURE 3.4.   Histograms of the distribution of the intensity ratings of 25 samples of potato chips for selected attributes.

ing attributes. Such an analysis is important because samples that do not fall at the extremes of any attribute ratings may still be unique in the sense that they fail to follow the pattern exhibited by the majority of the samples when two or more attributes are considered simultaneously. Such samples may be perceived as having an "unbalanced" or "unblended" flavor and, as a result, be less acceptable to consumers.

Scatterplots are good starting points for identifying relationships among the attributes and for singling out unique samples. Samples that do not follow the systematic pattern of the remaining samples should be set aside for further study. (The possibility of data entry errors should not be overlooked at this point, either.) The 153 scatterplots of the possible pairs of 18 attributes in the potato chip example are too numerous to present here, but that number of plots can be reviewed in a matter of minutes, to reveal the strength and nature of the pair-wise relationships among the attributes and to determine if any samples depart substantially from the general trends. In the example, samples 3 and 20 departed from otherwise linear relationships between saltiness and potato complex (see Fig. 3.5). Samples 3 and 20 are selected for consumer testing because of their unique behavior.

To summarize, after the first three steps of the example sample selection analysis:

1. Samples 4, 7, and 12 have been selected for consumer testing, based on extreme ratings in certain attributes.
2. Samples 3 and 20 have been selected for consumer testing because of unique patterns in their saltiness versus potato complex ratings.
3. Raw potato flavor, cardboard, and painty have been dropped from the list of 18 attributes that remained after step 1 (see Fig. 3.3), due to the lack of a variability that exists after one or more of the extreme samples (4, 7, and 12) were excluded.

STEP 4: Select Final Set of Samples for Consumer Testing.   In situations where only a small number of attributes are being considered, it is possible to make the final selection of samples for consumer testing directly from plots and tables of the descriptive attribute ratings. Samples should be selected to represent the low, middle, and high ranges of variability in each attribute. When the number of attributes is too large for direct selection of samples, principal components analysis (PCA) can be applied, to reduce the number of dimensions of product variability that need to be considered (see Appendix 4).

PCA applied to the 15 attributes that remain in the analysis after steps 1, 2, and 3 reveals that only two principal components (PCs) are required to explain 88 percent of the variability in the 20 remaining samples (i.e., less the 5 samples that

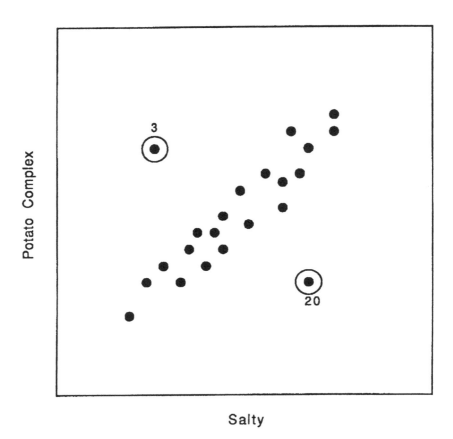

FIGURE 3.5.   Scatterplot of the attribute ratings of 20 potato chip samples for saltiness vs. potato complex intensity. Of the 20 samples, 18 exhibit a strong linear trend between the two attributes.

percent of the variability in the 20 remaining samples (i.e., less the 5 samples that were selected for testing in steps 2 and 3, earlier). The leveling out of the scree plot in Figure 3.6 shows that no significant gain in explained variability occurs by including additional components. The PC loadings of each attribute, in Table 3-6, are interpreted in the same way as correlation coefficients. The loadings can be studied to reveal the groupings of the attributes within each PC.

More important to our current purpose, the PC scores are computed for each sample using, for example, PROC SCORE (SAS 1989). The two PC scores per sample are much easier to examine than the original 15 attributes. Samples representing the low, middle, and high ranges of product variability can then be

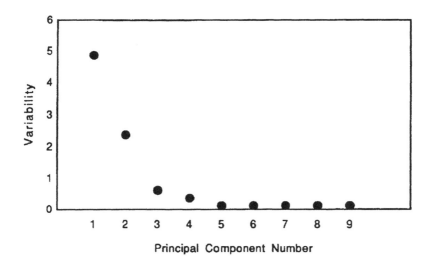

FIGURE 3.6.   A screen plot from the PCA (see Appendix 4), based on the descriptive attribute data from the potato chip example.

Table 3-6.   Principal component loadings of 15 sensory attributes for 20 samples of potato chips

Attribute	PC1	PC2
Color intensity	-0.8	0.2
Even color	0.7	0.2
Translucency	-0.1	0.9
Even size	0.6	0.3
Even shape	0.5	0.5
Folds	0.0	0.7
Potato complex	-0.8	0.3
Fried	-0.9	0.2
Salty	-0.8	0.5
Burn	0.4	-0.5
Oily surface	0.5	0.9
Hardness	0.2	-0.7
Crisp/crunch	0.7	-0.5
Denseness	0.6	-0.3
Toothpack	0.3	0.3

selected from this smaller number of dimensions. The samples tentatively selected for consumer testing are indicated in the plot of the two PC scores in Figure 3.7.

The last step before finalizing the sample selections is to examine the distribution of the tentatively selected set, along with those extreme and unique samples identified in steps 2 and 3, earlier, on their original attribute scales. Establish that the range of variability in each attribute is adequately spanned by the selected samples. If necessary, drop or add samples to achieve a uniform coverage of the observed ranges. For instance, in the example, none of the samples selected thus far had an extremely high rating in even color. Therefore, sample 8 is added to the

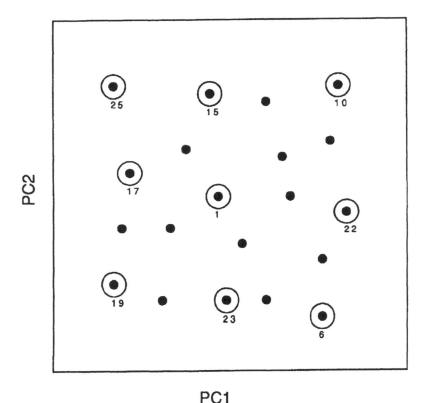

FIGURE 3.7.   A scatterplot of the principal component scores of 20 samples of potato chops. The highlighted samples were selected for consumer testing.

set of selected samples for no other reason than because sample 8 had the highest rating in even color.

Step 4 results in the selection of samples 1, 6, 8, 10, 15, 17,19, 22, 23, and 25 for consumer testing. Add to these the extreme samples 4, 7, and 12 and the unique samples 3 and 20, to form the complete set of 15 samples to be evaluated by consumers.

Planning and Execution of the Consumer Research Study
The consumer responses to the product's variability are collected by conducting a consumer research study. The methodology and the protocols followed in this study should be designed by an experienced sensory professional. This support is usually provided by the R&D sensory or product evaluation department. Alternatively, an outside consulting group can be used. With this support, the test design will include sound sensory procedures for the collection of valid and reliable consumer information.

Planning, administering, and executing consumer tests for QC purposes requires the involvement and assistance of many departments within a company. Among them are Operations/Manufacturing, Quality Control, Product Development, and Marketing/Market Research. The most important issues to consider when designing the consumer study are test location(s), consumer recruitment, test design, consumer questionnaire design, test execution, and validation.

Test Location(s).   The team of professionals working on the QC/sensory project needs to decide on the location or locations of the consumer test. The product's distribution areas and the company philosophy on regional versus national consistency play a role in this decision.

- Only one test location is needed when both the production and the product distribution areas are confined to one region.
- Several regions must be tested when there are manufacturing facilities in various locations, and the regional products differ due to either formula and/or processing differences.

The results from multiple locations can be used differently. Companies may treat data and programs regionally (establishing different specifications and QC/sensory programs in each region), or companies may try to standardize regional production differences prior to establishing specifications and the QC/sensory program.

Consumer Recruitment.   Medium or heavy users of the product should be selected. Interaction with the company's Market Research department is required at the planning phase, to select the screening criteria needed for the study.

Experimental and Test Design.   The consumer research needed in a QC/sensory project involves the evaluation of many samples. The sample selection process described above is geared toward reducing the total number of samples. However, a considerably large number of products is still tested. For example, for the potato chip example, a total of 15 products is to be consumer tested. The sensory professional in charge of the study selects the most appropriate design, based on the characteristics of the products and the total number of samples to be evaluated.

Unless the products require an in-home placement, a central location test (CLT) design is recommended. A CLT provides a better setup for evaluating a large number of samples. Situations that require an in-home placement are more expensive and time consuming than CLTs and, in some instances, may require that a less than optimal test design be used, in order to complete the study.

In selecting the most appropriate experimental design, the researcher has to decide on the use of a complete versus an incomplete block design (see Appendix 3). Typically, because of the large number of samples involved, an incomplete block design is used. In an incomplete block design, each respondent evaluates only a small number of the test samples. In a complete block design, each respondent evaluates all of the test samples. Choosing a complete block design involves either long evaluation sessions (e.g., two four-hour sessions) and/or multiple sessions conducted over several days, to be able to test all samples. These designs have advantages and disadvantages (Carr 1989) that should be considered when planning the study.

In the potato chip example, 15 samples were selected for consumer testing. To avoid respondent sensory fatigue and boredom, an incomplete block design was selected. In this design, each consumer evaluates only 3 of the 15 samples. Table 3-7 shows the characteristics of the design, including the total number of consumers needed to obtain the desired minimum of 70 replicate evaluations per sample.

Consumer Questionnaire Design.   The consumer questionnaire used in this study is designed to assure the collection of the type of data needed to relate the consumer information to the descriptive panel (intensity) data (see pg. 76). Figure 3.8 shows an example of the type of questionnaire used in the potato chip study. Other versions of this questionnaire could be used.

The questionnaire designed for this study should have two parts: acceptance and attribute questions. Figure 3.8 shows that, for the potato chip example, overall acceptance, appearance, flavor, and texture acceptance, and acceptance of specific attributes (potato flavor, crispness) are included. In addition, intensity questions for the same attributes are asked (e.g., potato flavor intensity, freshness intensity).

Other tests, such as descriptive and consumer qualitative tests (e.g., focus groups), are very useful when designing consumer questionnaires. The descriptive information is used to select the attributes to be included in the questionnaire. These attributes are those found to be the most variable in the product. For

Table 3-7.   Balanced incomplete block design used for consumer study of potato chips—15 samples evaluated in groups of 3

Repeat the design 10 times, using 350 respondents, to obtain 70 evaluations per sample.

Block	Samples		
(1)	1	2	3
(2)	4	8	12
(3)	5	10	15
(4)	6	11	13
(5)	7	9	14
(6)	1	4	5
(7)	2	8	10
(8)	3	13	14
(9)	6	9	15
(10)	7	11	12
(11)	1	6	7
(12)	2	9	11
(13)	3	2	15
(14)	4	10	14
(15)	5	8	13
(16)	1	8	9
(17)	2	13	15
(18)	3	4	7
(19)	5	11	14
(20)	6	10	12
(21)	1	10	11
(22)	2	12	14
(23)	3	5	6
(24)	4	9	13
(25)	7	8	15
(26)	1	12	13
(27)	2	5	7
(28)	3	9	10
(29)	4	11	15
(30)	6	8	14
(31)	1	14	15
(32)	2	4	6
(33)	3	8	11
(34)	5	9	12
(35)	7	10	13

example, the potato chip questionnaire includes the attributes color, evenness of color, potato flavor, saltiness, firmness, and crispness, because those are the most variable characteristics (Fig. 3.3). Focus groups, a qualitative consumer test, are useful in selecting the appropriate consumer terms to be included on the questionnaire. For example, "freshness" was selected with the information obtained in

NAME: _____

SET NUMBER: _____
SAMPLE NUMBER: _619_

POTATO CHIPS

Please look at and taste the product presented and indicate how much you liked/disliked it, considering all characteristics (Appearance, Flavor, and Texture).

DISLIKE   □ □ □ □ □ □ □ □ □   LIKE
EXTREMELY   EXTREMELY

Please indicate what you particularly liked or disliked about the product.

LIKED   DISLIKED

_____   _____
_____   _____
_____   _____

Please indicate how much you liked/disliked the following characteristics.

APPEARANCE
DISLIKE   □ □ □ □ □ □ □ □ □   LIKE
EXTREMELY   EXTREMELY

FLAVOR
DISLIKE   □ □ □ □ □ □ □ □ □   LIKE
EXTREMELY   EXTREMELY

TEXTURE
DISLIKE   □ □ □ □ □ □ □ □ □   LIKE
EXTREMELY   EXTREMELY

FIGURE 3.8.   Questionnaire used for consumer research study of potato chips.

SPECIFIC EVALUATION

Retaste the product as needed and mark your response for both questions (LIKING AND INTENSITY/LEVEL).

LIKING

INTENSITY/LEVEL

COLOR   dislike / like   light / dark

EVENESS OF COLOR   dislike / like   uneven / even

POTATO FLAVOR   dislike / like   none / strong

SALTINESS   dislike / like   not salty / very salty

FRESHNESS   dislike / like   not fresh / very fresh

FIRMNESS   dislike / like   not firm/soft / very firm

CRISPNESS   dislike / like   not crispy / very crispy

OILINESS   dislike / like   not oily / very oily

FIGURE 3.8.   (continued)

groups conducted for the potato chip project. Consumer responses to samples with a "cardboard" note (descriptive term) were "not fresh," "stale," and "old." The consumer term "freshness" was included, to capture the consumer responses to samples with a cardboard note.

Test Execution and Validation.   The sensory professional oversees the execution of the test, to ensure adherence to the test protocols and controls. As in other consumer tests, the actual execution of the test is a straightforward process, provided that the study has been carefully designed.

It is recommended that a validation of completed consumer responses be done. This practice involves contacting a subset of respondents by telephone, to confirm their participation in the test.

Data Analysis/Data Relationships

The relationship between the consumer responses and the descriptive panel intensity data are examined, to aid in establishing the QC specifications for each sensory attribute. The statistical methods applied to determine the nature of the relationships should be kept as simple as possible. Graphical methods may be all that are required.

While the relationship of the descriptive attributes to overall acceptability is most important, it is always informative to examine the relationship between the descriptive attributes and the acceptability ratings for specific attributes. In addition, an analysis of the relationship between the descriptive attributes and the consumers' intensity ratings of various product characteristics can be used to gain a clearer understanding of consumers' responses, their use of vocabulary, and so forth.

The first step in the analysis is to plot the overall acceptability and attribute acceptability responses versus the descriptive attribute ratings. A correlation analysis, run simultaneously, provides useful supporting information. Some descriptive attributes that fail to exhibit systematic relationships with overall acceptance may have a strong relationship with one or more of the attribute acceptance responses. The attribute acceptance relationships should not be ignored in setting QC specification for sensory responses.

The plots should be examined first to identify the attributes that exhibit no systematic relationship with either overall acceptance or any of the attribute acceptance responses (that is, plots that exhibit a completely random pattern, accompanied by a correlation coefficient near zero). The samples sent to the consumer test were selected to span the typical range of product variability; any attribute exhibiting a random pattern has no meaningful impact on acceptability (see Fig. 3.9a).

Next, interest should focus on those attributes that exhibit strong linear trends, either positive or negative, in overall and/or attribute acceptability versus descrip-

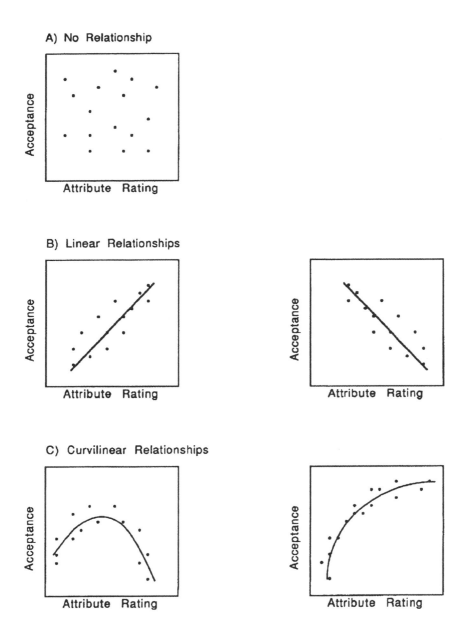

FIGURE 3.9.   Scatterplots of common relationships between acceptance responses (either overall or for attributes) and descriptive attribute ratings.

tive ratings. These trends are readily apparent in plots and are accompanied by large (positive or negative) correlation coefficients. A simple linear regression of the form

$$Acceptance = b_0 + b_1 (Attr)$$

where (Attr) is the descriptive attribute rating, can be used to obtain predictive equations that relate the value of a descriptive attribute to the consumer acceptability rating (see Fig. 3.9b). (In practice, an eyeball fit may be sufficient.)

Lastly, the plots of the descriptive ratings that exhibit curvilinear relationships with acceptability (both overall and for attributes) should be examined. For well behaved data an eyeball fit may be sufficient to develop a predictive relationship between the descriptive and acceptability responses, but, typically, a quadratic regression equation of the form

$$Acceptance = b_0 + b_1 (Attr) + b_2 (Attr)^2$$

is used (that is, both the descriptive rating and the squared value of the descriptive rating are used to predict the acceptability value (see Fig. 3.9c)).

The 18 descriptive attributes in the potato chip example are plotted versus the overall acceptability and the various attribute acceptability responses, using the 15 samples that were evaluated by both the descriptive panel and consumers. The following relationships are obtained (individual plots are not presented):

1. Translucency, even shape, folds, potato complex, raw potato flavor, burn, Oily surface and toothpack exhibit no systematic relationship with any of the consumer acceptability responses—neither overall nor any of the attributes (i.e., all plots resemble Figure 3.9a).
2. Even size, fried potato flavor, and crisp/crunch exhibited positive linear relationships with overall acceptance. Cardboard and Painty exhibit negative linear relationships with overall acceptance (as in Figure 3.9b). Denseness exhibits no systematic relationship with overall acceptance, but has a negative linear relationship with the consumer's airiness acceptability response (i.e., as in Figure 3.10).
3. Color intensity, even color, and saltiness do not exhibit a systematic pattern with overall acceptability. However, color intensity and even color exhibit a curvilinear relationship with the consumers' color acceptability response resembling the pattern presented in Figure 3.9c. In addition, saltiness exhibits a similar curvilinear relationship with the consumers' saltiness acceptability response. Only hardness exhibits a curvilinear relationship with overall acceptability.

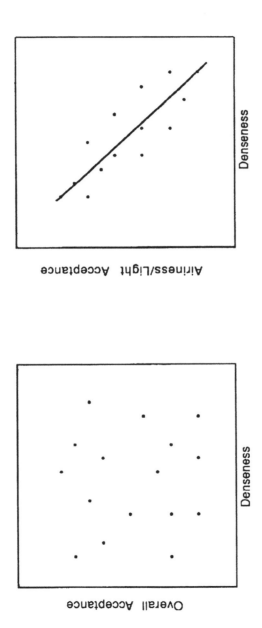

FIGURE 3.10. Scatterplots of overall acceptance versus denseness intensity (as measured by the descriptive panel) and airiness acceptance versus denseness intensity for the 15 potato chip samples.

In order for the relationships revealed in the analysis just discussed to be used to set QC specifications, management must select minimum product performance criteria for the overall acceptability response and, if necessary, for each of the attribute acceptability responses considered. No meaningful specifications can be set without this information because the descriptive attributes by themselves do not directly reflect acceptability. This issue is discussed in detail in the next section. In preparing for a management review meeting, the relationships between the descriptive and acceptability responses should be summarized using plots similar to those presented in Figure 3.9. Only those plots that clarify the relationship between the descriptive and consumer data should be presented.

Establishment of Final Specifications

A management meeting is scheduled to review the findings of the data analysis. It is important that upper management and all managers from each company's function (e.g., R&D, operations, QC/QA) attend. The purpose of the meeting is to establish the final sensory specifications for each attribute, by assessing the changes in the hedonic consumer's response to the product variability. The results are presented in two groups:

1. Those variable attributes which *did not show* an effect on consumer acceptance (Fig. 3.9a); and
2. Those variable attributes which *did show* an effect on consumer acceptance (Fig. 3.9b and 3.9c).

The discussion and the decisions made are completed by examining one attribute at a time. For each attribute, the following is reviewed:

- The definition;
- The demonstration of the variability by actual production samples (the lowest, the highest, and one or two intermediate intensities are shown); and
- The plots displaying the data relationships between consumer responses and panel attribute results.

Attributes showing a relationship depicted in Figure 3.9a (e.g., translucency, folds) did not show an effect on consumer acceptance. For those attributes, no specification needs to be set, since consumer's acceptance is not affected by its variability. Therefore, these are attributes that need not be measured by the panel. Despite this, some companies may decide to have the panel measure one or two attributes from this category. This is done either for documentation purposes or for tracking attributes that are of particular interest to the company.

Attributes with relationships as those shown in Figure 3.9b (e.g., crispness) and

Figure 3.9c (e.g., saltiness) affected consumer acceptance and are used to set specifications. The final specifications represent the limits of the variability range tolerated for each attribute, expressed in terms of descriptive intensities.

In order to set the QC specifications on the attribute ratings, minimum product performance criteria must be established for the overall acceptability response and each of the critical attribute acceptability responses. No meaningful specifications can be set without this information because the descriptive attributes by themselves do not directly reflect acceptability. Management of the potato chip company has established an overall acceptability of 6.5 (on a liking scale from 1-9) to be the minimum product performance criterion. This cutoff point is used to set specifications.

Figure 3.11 illustrates this process. With a cutoff point of 6.5 in overall acceptance, the specification set is 5.8-10.0. Products with an attribute intensity lower than 5.8 or higher than 10.0 yield a consumer acceptance score lower than

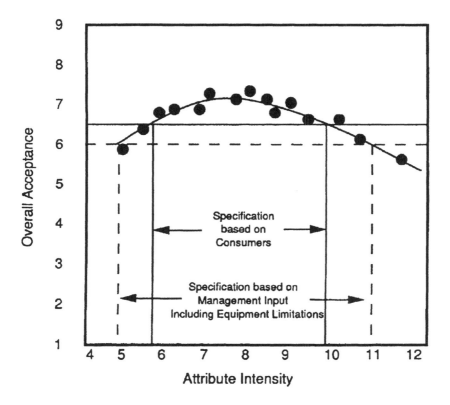

FIGURE 3.11.   Establishment of specifications, a) based on consumer responses and b) based on consumer responses and other issues, such as equipment limitations.

the minimum established (6.5) and are therefore to be considered unacceptable (out-of-specification).

Other factors beside the consumer input are taken into consideration when setting specifications, such as production limitations, costs, and management quality expectations. This means that very rarely are the specifications set based on the straightforward inspection of the relationships, as discussed earlier.

In reality, the final specifications might be set with either a narrower or a wider intensity range than that determined by the acceptance cutoff point. Narrower ranges are set by companies that want stricter criteria than the consumers themselves. Wider ranges than those called for by consumers may be set by companies for which the cost to maintain the narrower specifications is too high, compared to the benefits gained. For example, very few companies would incur the high cost of upgrading all their processing equipment, to make sure that 100 percent of the production falls within a tight specification range. These concepts are also illustrated in Figure 3.11. The manufacturer of potato chips set a consumer acceptance cutoff point of 6.5 or higher. Based on this criterion, the specification for the attribute displayed in Figure 3.11 was set at 5.8 to 10.0. However, it is impossible to have more than 80% of their production fall within the range determined and achieve the cutoff point of 6.5 in acceptance with the current processing conditions. In that case, instead of setting the specifications for that attribute as 5.8 to 10.0, it is set at 5.0 to 11.0 (Fig. 3.11).

Table 3-8 shows the final specifications set for the attributes measured for potato chips.

Table 3-8.   Final sensory specifications for potato chips

Attribute	Score
APPEARANCE	
Color intensity	3.5–6.0
Even color	6.0–12.0
Even size	4.0–8.5
FLAVOR	
Fried potato	3.0–5.0
Cardboard	0.0–1.5
Painty	0.0–1.0
Salty	8.0–12.5
TEXTURE	
Hardness	6.0–9.5
Crisp/crunch	10.0–15.0
Denseness	7.0–10.0

Specifications Set Based on Management Criteria Only

The involved process required in collecting consumer responses to the product's variability forces some companies to choose the more simple method for setting specifications. Although the authors prefer the more sound procedure (using consumer input) and consider that to be the more appropriate process of the two, a brief review of the alternative is discussed.

When specifications are set based on management's criteria only, the completion of a product survey and the summary of the results are needed. This means completing only two or three of the seven steps otherwise required when a consumer research study is undertaken.

The specifications without the consumer input are also set in a management meeting. In this meeting, the following aspects are presented:

- The results of the product survey completed for the identification of variability;
- A subset of production samples that demonstrate the range of the most variable attributes.

The results of the product survey can be summarized as in Table 3–5 and Figure 3.3. This figure shows the type of variability determined by the survey. Attention is focused on the attributes with the largest variability (asterisks in Fig. 3.12) and on off-notes (or other attributes where the variability range might be small but important).

The subset of production samples shown to management are the products that represent the extremes of the variable attributes and some intermediate points. The 3 to 4 samples shown to management for each variable attribute are marked with circles in Figure 3.12. The inspection of samples is essential in this decision-making process, since management has to understand both product attributes and their intensities.

The sensory specification for each variable attribute is set by management, after the inspection of the samples shown to demonstrate each variable attribute. Based on *their criteria only*, management decides how wide or narrow each specification (tolerable variability range) should be. For example, in setting specifications for cardboard and denseness, management reviewed the corresponding samples (Fig. 3.12) and set the following specifications:

Attribute	Range of Variability Observed	Specification
Cardboard	0–5.0	0.0–3.5
Denseness	6.0–10.0	6.0–10.0

These specifications are used in the program as described in later sections of this chapter.

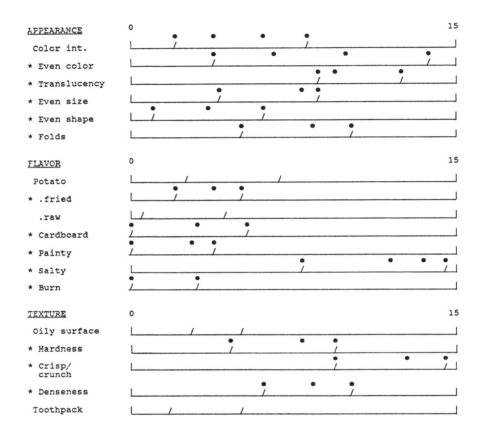

FIGURE 3.12.   Production variability of potato chips and products used for setting specifications through mangement criteria.

Training and Operation of Plant Panel(s)

A well-trained panel needs to be established at the plant(s) where the sensory characteristics of daily production will be measured and controlled. The panel can be formed either with the participation of company employees or of local residents. There are advantages and disadvantages in these two approaches (Mastrian 1985). The advantages of having local residents as panelists include the unrestricted time availability for panel sessions, the reliability of panel attendance, and the reduced overhead costs; disadvantages include the additional time required to train nontechnical personnel or people with little or no experience with the company's products, and the problems involved in handling legal and/or confidential issues when using nonemployees. The advantages of having company employees as panelists include their proximity to the evaluation location and the use of their technical expertise and/or product experience in the program. However, the main advantage in using employees in the QC/sensory program is that the product quality message is disseminated throughout the plant (Keeper 1985). The disadvantages of using employees on the panel include the logistical problems of adjusting production schedules to allow for the release of line workers, the lack of motivation and support encountered in the long run by plant supervisors, and the resulting lower motivation from the panelists themselves.

Training Steps

The panel is established through three phases: prescreening, screening, and training. The description of these three phases given below corresponds to the process followed in the establishment of a Spectrum panel (Meilgaard, Civille, and Carr 1987).

Prescreening

In the prescreening stage, potential panelists are contacted via a prescreening questionnaire to determine their interest and availability for participation. If employees are being screened, an indication of management support and interest in the program (e.g., a letter from the plant manager) is helpful in obtaining cooperation at this stage. Optimally 100 to 120 people are initially contacted at the prescreening phase, in order to select 50 to 60 people for the screening phase. However, for small manufacturing facilities, the number of panelists initially screened may be lower. In order to be selected for this phase, potential panelists must meet the following criteria: no allergies, availability, limited traveling, interest, ability to describe simple sensory characteristics, and the ability to use a scaling system.

Screening

Participants who are potential candidates, as indicated by the prescreening results, are scheduled to participate in a series of acuity tests and an interview (50 to 60 people). The standard tests and procedures used to screen QC/sensory panelists are modifications of the procedures used to develop an R&D descriptive panel (Meilgaard, Civille, and Carr 1987). These modifications include 1) the use of a smaller set of samples/products for the descriptive tests (e.g., odorants to describe aromatics) and 2) the use of actual production samples that show product differences in the detection tests. The purpose of these tests is to select the panelists who have normal acuity, are able to detect differences in production samples, and show some ability to describe sensory characteristics and to learn a scaling system.

Training

Optimally, a total of 20 to 30 participants are selected to participate in the training program. This large pool is recommended to assure a sufficient number of panelists per evaluation (e.g., 10 panelists). Plants that operate multiple shifts may conduct at least two training sessions to include the personnel from all shifts. A smaller number of panelists is selected in small plants, due to fewer staff members working at the production facility. In that case, a more involved training program is required to obtain a low within panel variability, which is achieved through additional training and practice time.

Depending on the product category and the number of attributes, the training program requires two to five 6-hour training sessions, for a total of 12 to 30 hours. Additional sessions are scheduled thereafter for practice.

Prior to the initiation of the training program, a plant sensory coordinator must be identified. The plant sensory coordinator gets involved in all the steps of the training, practice sessions, and panel maintenance. This participation and his or her interaction with the R&D sensory professional or consultant are required to assure the coordinator's independent work in later stages of the program.

During training, it is important that the sensory coordinator participates as a panelist trainee, *not* as a technician. The coordinator needs to have at least the same panelist skills as the panel itself, so the sensory coordinator can monitor the panel performance, understand panelists' and product problems, maintain the panel, and operate the program.

The most important component of the training is the product training phase, where the panel learns to evaluate the variable product characteristics for which sensory specifications were set.

For the potato chip example, the plant panel is to be trained to evaluate the following characteristics, which were found to have an impact on consumer acceptance:

- Color intensity
- Evenness of color
- Evenness of size
- Fried potato
- Cardboard

- Painty
- Salty
- Hardness
- Crispness/crunchiness
- Denseness

The "Training Program" section on page 88 presents the details of the training program.

Sample Selection for Training

For products with a short shelf life, a second survey is conducted to obtain the training samples. Samples selected in previous surveys and used to complete prior tests may have changed by the time the training is scheduled. At this point of the program, enough information and product experience has been gained to be able to conduct a smaller yet effective survey. Knowing the attributes and attribute intensities that need to be represented by products, the survey becomes simpler. The process then consists of finding the production samples that represent the various attributes and intensities needed for training. For products with a long shelf life, the samples needed for training may be available from previous surveys, if enough product has been stored.

Optimally, at least three intensity references per attribute are selected and shown to panelists during training. Figure 3.13 shows the training samples identified for the potato chip program marked with circles. The most important samples for training purposes are those that display the highest and lowest attribute intensities. It should be noted that certain extreme intensities (low or high) found in previous surveys and even tested in the consumer study might not be found in the survey that is completed prior to the training. Figure 3.13 shows this for evenness of size and cardboard. This might pose some difficulties in the panel training. The panel could accurately rate higher or lower intensities than those demonstrated in the training, because of the use of both internal references (e.g., specific product references like potato chip references shown during training) and external references (Muñoz 1986; Meilgaard, Civille, and Carr 1987). However, it is recommended that further production sample screening be done to find those products

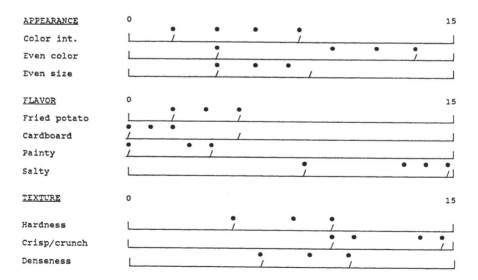

FIGURE 3.13.   Training samples (·) selected for the QC/sensory program of potato chips.

for the panel, even shortly after the panel has been trained. Alternatively, accelerated shelf life or other product stress procedures or R&D-created samples might be considered to provide those products needed for training. This might be required to obtain the high intensity of cardboard (5.0) found in previous surveys (Fig. 3.13), but not in the one completed for training.

Training Program

The panel training program should be conducted by an experienced sensory professional. The two main components of this training are the basic sensory phase and the product training phase. The objective of the basic sensory program is to familiarize panelists with the applications of sensory evaluation, its importance within the company and in the measurement of quality, and the basic concepts of physiology that play a role in the evaluation of the company's product characteristics (e.g., appearance, flavor, skinfeel).

In the product training phase, each of the attributes, for which sensory specifications have been set, is discussed one at a time. The production samples, representing ranges for each attribute that were previously identified, are presented to the panelists. The definition, evaluation procedures, and scoring of each attribute are covered, using external and internal (production samples) references. Table 3-9 shows the process for the training of denseness for potato chips. The same procedure is followed for all attributes.

Table 3-9.   Example of training process for the descriptive evaluation of production samples

Denseness

Step 1

Definition: Compactness of the cross section of the sample after biting completely through with the molars.

Low————->High

(airy)   (dense/compact)

Step 2 (optional)

Presentation of external (not product specific) intensity references*

Nougat (Three Musketeers—M&M Mars)	Denseness = 4.0
Malted Milk Balls (Whoppers)	Denseness = 6.0
Fruit Jellies (Chuckles—Nabisco)	Denseness = 13.0

Step 3

Presentation of internal references (potato chips production samples) (from Fig. 3.13)

Sample 1—Denseness = 6.1
Sample 2—Denseness = 8.5
Sample 3—Denseness = 10.0

*Muñoz (1986)

After this review, test production samples are presented for evaluation. These test samples are only coded with three digit codes and panelists are asked to rate the intensity of the attribute they just learned. Initially, the samples representing the largest difference in an attribute are presented as test samples. For example, for the saltiness exercise for potato chips, the most different samples available, representing the lowest (8.0) and highest (14.5) intensity in saltiness are chosen for the first panel exercise. Panelists rate these two samples for saltiness, and the results are compared to the documented saltiness intensities of those two samples (8.0 and 14.5).

In the two to five training sessions, all variable attributes are reviewed and the product evaluation exercises are completed. Thereafter, practice sessions are conducted to reinforce the concepts learned and give the panelists additional practice in the product evaluation. In each practice session, each panelist is asked to evaluate selected production samples, presented as unknowns, on a daily basis. Records of the evaluations are kept and used for panel feedback. Production samples identified in the survey are used for the practice sessions. Therefore, descriptive evaluations are available and are necessary for judging panel performance. Table 3-10 shows the process for the potato chip training. The information listed under "reference" corresponds to the descriptive characterization of the production sample by the trained and experienced panel (like an R&D or contract research panel). This information was collected prior to the start-up of the training. The information listed under "plant panel" is the average rating of the plant panel (n=10) of the same production sample. Simple inspection of these data, and similar information collected across all practice days for a variety of products, indicates the performance of the panel. In addition, the individual observations are assessed to conclude on each panelist's performance. Table 3-10 shows that the panel rated the saltiness intensity lower and the intensity of fried potato higher than it should have. If this behavior is observed consistently over the practice sessions, these two attributes should be reviewed with the panel. Product references and other training material are used for this review.

Once the practice period is completed and problem areas have been reviewed and resolved with the panel, the ongoing operation of the program can be started. This would mean that the panel is qualified to accurately evaluate actual production samples in terms of their variable sensory attributes. Furthermore, the panel's data can then be handled with other QC data for product decisions.

Panel Maintenance
As with any other sensory program, a panel maintenance program needs to be established. This maintenance program should address both the psychological and physiological factors of the panel.

Table 3-10. Example of panel performance using descriptive results (potato chips)

	Intensities	
	Expert Panel Reference	Plant Panel
APPEARANCE		
Color intensity	5.4	5.0
Evenness of color	12.1	12.7
Evenness of size	7.3	8.5
FLAVOR		
Fried potato	5.2	7.1
Cardboard	0.4	0.0
Painty	0.0	0.0
Salty	13.6	8.3
TEXTURE		
Hardness	8.7	8.3
Crispness/crunchiness	14.1	13.5
Denseness	8.5	7.7

Psychological Factors

The psychological factors are, in the long run, the most important factors that affect panel performance. A QC/sensory panel that frequently participates in routine evaluation of production samples experiences boredom and lack of motivation after a period of continuous panel participation. This problem might be further aggravated by other factors, such as lack of support from immediate supervisors, lack of support for relief from the line, plant layoffs, cancellation of panel sessions because of plant crises or production schedules, union problems, boredom in the evaluation process because products are too consistent, and so on.

The plant sensory coordinator should administer a program to maintain the panel's interest, motivation, and, consequently, adequate performance. Some possible activities are:

- Recognizing and appreciating the panel participation by the sensory coordinator, the QC manager, the panelists' supervisors, and, most importantly, the plant manager. The panelists need to be reminded on a regular basis of the importance of their participation in the program.
- Acknowledging the panel participation in solving plant problems and/or the panel support in special plant programs or efforts. This point is related to overall recognition. Tangible results of critical panel contribution to the plant's opera-

tion should be shared not only with the panel, but also with the personnel of the whole plant.

- Planning special group activities outside the working place. These group activities foster interaction and friendship among panel members.
- Giving small rewards to the panel on an occasional basis. Every company differs in what kind of small rewards are given to the panel for their participation. Some include trinkets, T-shirts, company products, other products, tokens, gift certificates, lunches, dinners, and so forth.
- Scheduling panel review sessions with the whole panel (or at least groups of ten people) to review the technical aspects of the program (e.g., attributes, references). The panel then realizes that the program is important in the company and deserves attention and maintenance.
- Providing panel performance feedback. The sensory coordinator informs the panel on a regular basis of the individual and whole panel performance. This is possible by the use of blind controls presented in regular evaluations. Both the adequate performance and problem areas are shared with the panel.
- Recognizing individual panelist's participation. Notable contributions of panelists (attendance, correct identification of blind controls or off-notes,) should receive recognition by the sensory coordinator and the QC manager.

There are many unique ways by which a company can acknowledge the panel participation. The philosophy and operation of each company determines the most effective activities and aspects of this program. One factor to consider for in-plant programs is the restrictions that plant unions can pose on these activities. In extreme cases, no tangible rewards can be given to any of the panelists. In such companies, the sensory coordinator has to rely more heavily on the psychological rewards to the panel for maintaining their interest and motivation.

Physiological Factors
As with the psychological maintenance program, there are numerous activities that can be administered by the sensory coordinator to maintain the panel's technical performance. Among the most common and recommended ones are:

- Scheduling review sessions. The panel coordinator meets either with the whole panel or groups of ten panelists. In these sessions, the technical aspects of the program are reviewed: attribute definitions, references, and evaluation procedures. Other aspects of the program are discussed, such as scheduling, environmental effects on evaluations, testing controls, and other issues that might have an effect on panel performance. These sessions should be scheduled regularly (perhaps once a month). Their frequency is to be determined by the panel and sensory coordinator, depending on their needs.

- Reviewing intensity references on a regular basis. Although this should also be done in the panel review sessions, the review of intensity references should be an ongoing procedure. Once a week or every other week, the sensory coordinator prepares a product reference (internal reference) and gives it to the panel with a "completed" scoresheet that lists the intensity values for all attributes. These intensities are obtained either from a more experienced panel (such as an R&D panel) or are the average intensities of the plant panel from the evaluation conducted one or two days before. This serves as a review of attributes evaluated and, more importantly, of the attribute intensities.
- Monitoring panel performance through the evaluation of blind references. Blind controls or references should be given to the panel in the regular evaluation sessions on a routine basis. The descriptive characterization of these references should be available for data comparison. This characterization is obtained from an R&D panel or from the long-term average results of the plant panel itself. Panel performance is monitored by following the same procedures used in the practice period at the end of training (Table 3-10). Comparing the reference results and the average plant panel data allows the sensory coordinator to discover problem areas in the evaluation of product attributes.

In addition, the difference between the plant panel's average result and the reference value for each attribute can be plotted on an I-chart, as in Figure 3.14, to track the consistency of the plant panels ratings over time.

- Monitoring individual panelist performance. The performance of each panelist should be monitored in terms of the following:
    - Individual measurement compared to the panel mean;
    - Individual measurement compared to the results of the blind control; and
    - Individual performance per attribute over sessions.

Calculations of the within and between panel variability per attribute (see Appendix 5) are useful to monitor panelist-to-panelist and session-to-session variability, respectively. In addition, I- and R-charts can be generated for panel performance. For example, the range of the individual panelists' ratings can be plotted on an R-chart (Fig. 3.15) to assess panelist-to-panelist consistency. I-charts similar to Figure 3.14 can be generated for each panelist by plotting the difference between the individual's rating and the panel mean and/or the reference value each time the hidden reference sample is evaluated. Panelists with consistent positive or negative biases or erratic, highly variable ratings are clearly evident.

- Training on the evaluation of other products. In an established QC/sensory program of a company manufacturing a variety of products, the training on other products is completed on a continuous basis. Training the panel on other

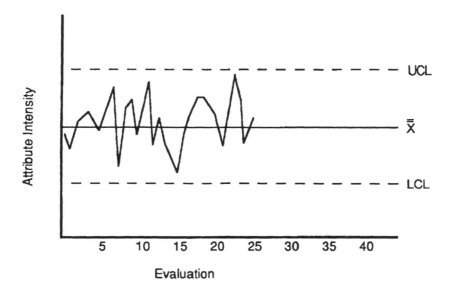

FIGURE 3.14.   I-chart of the plant panel's rating of a representative attribute for a hidden reference sample presented routinely during the operation of the program.

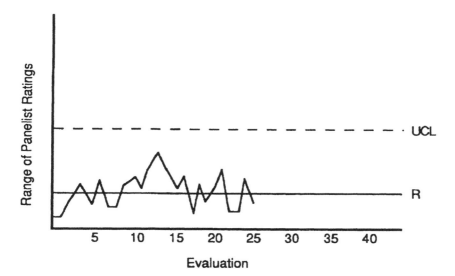

FIGURE 3.15.   R-chart of the plant panel's rating of a representative attribute for a hidden reference sample presented routinely during the operation of the program.

products implies broadening the panel skills or improving the existing ones. With the introduction of new products to the program, the panel reviews and improves on the attributes already learned and learns new attributes.

Establishment of a Data Collection, Analysis, and Report System

Data Collection

Samples to be evaluated by the plant panel are collected at the same time as the samples that will undergo other QC testing. The frequency of sampling is determined using the same QC criteria that apply to all other tests. For the potato chip example, each lot of finished product is sampled three times—early, middle, and late in the production run—just prior to packaging.

The number of samples submitted to the panel per session affects the sample presentation design used by the sensory coordinator. If the number of samples per session is small, a complete block presentation design is used. If, however, the number of samples per session exceeds that which can be evaluated before sensory fatigue sets in, then a balanced incomplete block presentation design is used (see Appendix 3). A complete block presentation is used for the potato chip samples.

Data Analysis

For each attribute, the individual panelist's ratings are recorded. Table 3-11 contains the panelists' perceived saltiness ratings for the three samples of potato chips collected during routine QC sampling. Note that in Table 3-11 the scores for the three samples (early, middle, and late) are recorded separately, so that

Table 3-11.   Panelist Data from QC/sensory Comprehensive Descriptive Program

Attribute: Saltiness			
		Sample From Lot 011592A2	
Panelist	Early	Middle	Late
1	10.9	11.8	11.8
2	10.6	12.4	12.4
3	10.5	11.8	11.5
4	10.6	12.3	13.0
5	10.8	11.0	12.3
6	10.4	12.2	11.5
7	10.4	11.8	13.6
8	10.1	11.2	13.0
9	10.5	12.2	11.6
10	10.0	11.6	12.3
Mean	10.5	11.8	12.3

within-lot variability can be tracked. The panel's average attribute rat-ings for each of the samples, which appear in the last row of Table 3-11, are the raw QC/sensory data associated with each lot of product. For complete block presentation designs, arithmetic means are used. For incomplete block presentation designs, adjusted (or least square) means must be computed (Cochran and Cox 1957).

If multiple samples per lot are collected, as is done in the potato chip example, the lot average and range (i.e., the average and range of the panel means) are computed and recorded (see Table 3-12). If only a single sample per lot is collected, the panel mean is the mean rating for the entire lot; no measure of within-lot variability is available. Note in Table 3-12 that lot # 011592A2 is out of specification by being too low in "even color" and too high in "card-board."

The potato chip manufacturer has an SPC program in place, so the sensory coordinator maintains the control charts of the sensory attributes that are measured on the finished products. If multiple samples per lot are collected, as is done for the potato chips, $\overline{X}$-Charts and R-charts are used to track production. If only a single sample is collected per lot, I-charts are used in place of $\overline{X}$-charts. No analogue for R-charts is available when only one sample per lot is collected.

Figure 3.16 presents the $\overline{X}$-chart and R-chart for the "color intensity" attribute for potato chips. Included in the $\overline{X}$-chart are the process average ($\overline{\overline{X}}$ and 3σ control limits (UCL and LCL) (see Appendix 5). The process average and control limits are based on the historical performance of the process. Using the results of the

Table 3-12.   Summary Analysis of QC/sensory Attributes

| Attribute | Sample from Lot   011592A2 | | | Mean | Range |
	Early	Middle	Late		
Color Intensity	4.6	4.2	5.0	4.6	0.8
Even Color	4.8	4.6	4.2	4.5*	0.6
Even Size	4.1	5.2	6.2	5.2	2.1
Fried Potato	3.6	3.3	3.9	3.6	0.6
Cardboard	5.0	3.3	4.6	4.3**	1.7
Painty	0.0	0.0	0.0	0.0	0.0
Salty	10.5	11.8	12.3	11.5	1.8
Hardness	7.0	7.5	7.9	7.5	0.9
Crisp/Crunch	12.5	13.1	14.0	13.2	1.5
Denseness	7.4	7.1	7.5	7.3	0.4

* = Below Low Specification Limit
** = Above Upper Specification Limit

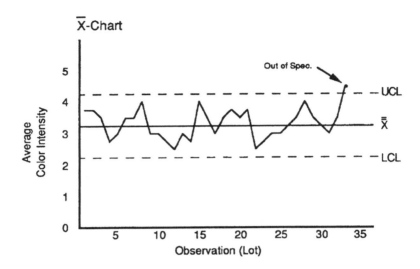

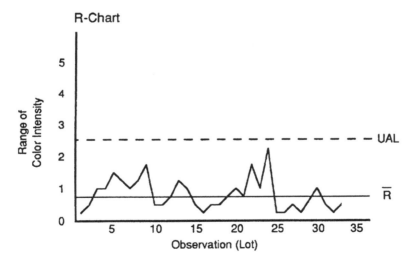

FIGURE 3.16.   X-chart and R-chart of the color intensity of potato chips showing that the latest lot evaluated, although "in-spec," falls above the upper control limnit (UCL), indicating that the process is out of control.

initial product survey (Table 3-5), the long-term average color intensity is 3.2. The control limits for color intensity are computed as

$$UCL = \overline{\overline{X}} + 3\sigma/\sqrt{n}$$
$$= 3.2 + 3(0.5)/\sqrt{3}$$
$$= 4.1$$

and

$$LCL = \overline{\overline{X}} - 3\sigma/\sqrt{n}$$
$$= 3.2 - 3(0.5)/\sqrt{3}$$
$$= 2.3$$

where σ is the long-term standard deviation of color intensity (from Table 3-5) and n is the number of samples collected per lot during routine QC sampling (i.e., n = 3; early, middle, and late in the lot). The $\overline{X}$-chart shows that, although color intensity is in-spec (Table 3-8), the value for the latest lot (#011592A2) lies above the upper control limit. Over and above the out-of-spec conditions previously noted, this further indicates that the process is out of control and that corrective actions need to be taken.

The control limits for the R-chart are established by multiplying the average range of intensities of the three QC samples (e.g., $\overline{R}$ = 0.85 for color intensity, based on historical data) by the entries corresponding to n = 3 in Table A-10 at the end of the book. The 3σ "action limits" for color intensity are

$$LAL = 0.04(0.85) = 0.0$$

and

$$UAL = 2.99(0.85) = 2.5$$

There is no indication of excessive within-lot variability in color intensity for Lot #011592A2 of the potato chip example.

The average value for a lot does not have to fall beyond the control limits in an $\overline{X}$-chart in order to justify the conclusion that the process is out of control. Figure 3.17 shows the $\overline{X}$-chart for the hardness attribute of the potato chips. The most recently observed lot, with an average value of 7.5, is the ninth consecutive lot with a hardness rating above the long-term average of 7.3 (Table 3-5). Based on Nelson's (1984) criteria (see Appendix 5), this occurrence is another signal that the process is out of control, even though the most recently observed lot has a hardness rating very close to the long-term average.

Report System
The sensory coordinator should highlight any samples that fall out of specification, using the company's standard procedures for such occurrences as presented, for example, in Table 3-12. Additionally, if an SPC program is in effect, the control charts for each of the attributes should be updated each time a new lot of product is evaluated as

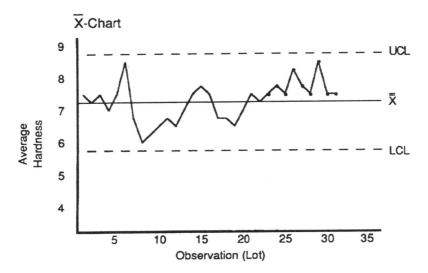

FIGURE 3.17. $\overline{X}$-chart of the hardness of potato chips showing that the latest lot evaluated is the ninth consecutive lot to fall above the process mean ($\overline{X}$), indcating that the process is out of control.

in Figures 3.16 and 3.17. The updated charts should then be forwarded to appropriate personnel. The key factor to bear in mind is that the QC/sensory data should be treated in exactly the same manner as any other analytical measurements being collected for QC. All standard company formats, reporting procedures, and action standards apply. Once in operation, a successful QC/sensory function should be a fully integrated component of the overall QC program.

Characteristics of the Ongoing Program

After completing all the steps described in this chapter, the routine program can be established and operated on an ongoing basis. All QC/sensory programs consist of daily/routine and long-term activities.

Routine Activities
Scheduling of Panel Sessions
At the start-up of the program, the sensory coordinator spends a considerable amount of time scheduling and conducting the panel sessions for a given week. At this time, many program components have yet to be established, such as the sampling schedule (product collection), panelist's individual schedules, and procedures for relieving line workers. Once all the program parameters have been delineated, routine evaluations can start.

The routine schedule of panel sessions is another factor unique to each company.

Some companies might decide to schedule only one session per day to evaluate production samples from the previous or same day. Others might decide to schedule more than one session per day, since many more products are to be evaluated. This decision has to be made in the preliminary and planning phase of the QC/sensory program because of its impact on other characteristics of the program.

Based on the product sampling plan chosen (which determines the total number of products to be routinely evaluated), the product characteristics (which determine the maximum number of samples to be evaluated in a session), and the maximum time a panelist can spend in panel session, the following are determined:

- The number of panel sessions scheduled in a day;
- The times (schedule) of the those sessions; and
- The logistical requirements for the completion of the evaluations, besides personnel (i.e., the product pickup, product delivery to sensory area, materials needed, use of evaluation room, etc.).

These aspects should be established and then become a routine in the operation of a QC/sensory program.

Scheduling of Panelists' Participation

Careful assessment of the program's needs and resources at the planning phase of the program assures the use of a sufficiently large pool of panelists per evaluation (e.g., ten panelists) and avoids the continued routine participation of panelists without a recess.

In the authors' experience, scheduling panelists in daily evaluations in one week, every other week, is an adequate schedule for good panel performance. Participation can possibly be increased to weekly evaluations (with three to four days per week) if several product types are evaluated. This is possible, since the product diversity introduces some variety to the routine evaluation sessions of a QC/sensory program. However, as more products get integrated into the QC/sensory program, consideration should be given to training new panelists, rather than using the same panelists to participate in daily evaluation sessions. The busiest evaluation schedule a panelist should be given should be two sessions per day on a routine basis. Even then, this situation is only effective when a variety of product types are evaluated throughout the daily sessions.

Considering the above factors, each panelist's schedule should be prepared at the start-up of the program and communicated to both the panelist and his supervisor. Ideally, this schedule can be maintained for at least a month. Realistically, however, due to panelists' sickness, unexpected unavailability, and so on, a new schedule may have to be prepared on a weekly basis by the sensory coordinator.

In scheduling each panelist's series of evaluations, it is important that the following be observed:

- The schedule and/or its modifications must be communicated to both the panelist and his supervisor.
- Early notification assures the panelists' availability (especially in the case of line workers).
- Each panelist's schedule has to conform with the maximum number of sessions allotted per panelist in a week and the maximum time allowed per session.

Administration of Test
In administering each test, the sensory coordinator must give attention to:

- Adherence to adequate experimental and test designs;
- Administration of product pickups;
- Sample preparation and presentation;
- Preparation of references;
- Preparation of scoresheets or data collection system;
- Use of adequate evaluation procedures;
- Maintenance of test controls; and
- Complete and adequate data collection.

The time spent in the administration of the evaluation sessions by the sensory coordinator depends on the type of product evaluated, the evaluation procedures, and the data collection system of the program. The most involved programs are those requiring very controlled and complex product preparation procedures. Among these include certain food products requiring cooking or heating. In such cases, the program requires not only the complete supervision of the sensory coordinator, but the help of one or more QC/sensory technicians. When complex sample preparation is required, a more structured scheduling of sessions and panelists is necessary to ensure the integrity of the samples evaluated.

Conversely, other programs require minimal supervision. An example is a program for the evaluation of hard candy or paper products. As long as the facility offers the required environmental controls, panelists can complete their evaluation at the most convenient times for them. Products can be prepared in advance and left in the evaluation room for each panelist. The sensory coordinator requires minimal technician assistance and he or she is only needed to answer special questions or solve problems.

Data Analysis, Report, and Integration with Other QC Data
The QC/sensory data should be reported in the same manner as any other QC data. Timing and format are established based on current company QC practices.

Scheduling of Special Sessions
From time to time, special sessions are scheduled at the request of the plant or quality control managers to collect sensory information on production batches identified with problems or special characteristics.

These sessions usually take priority over all other activities of the sensory program, since they are critical for the plant's operation and also for the growth of the sensory program within the plant. The extent to which an established sensory program can assist in solving production problems will determine the degree to which the sensory program is recognized and supported.

The sensory coordinator should welcome those situations and provide full support. Furthermore, the sensory coordinator should be involved in all problem-solving situations for the benefit of the program.

Panel Maintenance
The panel monitoring and maintenance should be an ongoing and important activity of the sensory program. Several procedures have been discussed earlier. In addition, any other procedure used for panel monitoring and found in the sensory literature should be considered in QC/sensory programs (Amerine, Pangborn, and Roessler 1965).

Long-Term Activities
The long-term activities involve the growth and extension of the QC/sensory program and include activities related to program, panelists, and the sensory coordinator.

Program Issues
The established sensory program can grow through the introduction of other products to the program, the extension of the program to other production phases, and the involvement in plant-based research projects.

Additional Products.   The support, growth, and success of a QC/sensory program is evident by the degree and frequency with which additional products are incorporated into the program. As this occurs, the sensory program further develops in several ways. New panels are trained, panelists' skills improve, the sensory staff expands, the sensory facility might be remodeled or expanded, and the program becomes an essential component of the plant's QC program and operation.

Evaluation of New Production Phases.   As the QC/sensory program grows, other points of evaluation in the process are incorporated into the program. Those programs dealing exclusively with finished products are expanded to include the evaluation of raw or in-process materials. Conversely, those programs established for the evaluation of raw ingredients expand into the assessment of finished products.

The ultimate goal of an in-plant program is to cover all critical production facets of all the important product categories. Since all production facets for a variety of products cannot be covered by a program, the resources should be allotted to those areas yielding the maximum benefits and savings for the plant. This is the situation of a program whose efforts and resources focus on the evaluation of raw materials, with occasional evaluation of in-process and finished products as final controls.

Research Projects.   A variety of research projects need to be undertaken by the sensory coordinator for the program's growth. These projects are generally completed in collaboration with the R&D sensory and other departments.

There are many areas of research within a QC/sensory program. The priority placed in these projects is company dependent. Realistically, there are many companies that will never undertake a research project, since their budgets and resources are limited to the routine administration of the in-plant program. For others, such research may be an essential component of the program.

There are two main research areas of importance for the QC/sensory program's growth:

- The assessment of sensory instrumental relationships; and
- The evaluation of consumer response to product changes and variability.

Both areas are geared to reduce the routine tasks and work load of the sensory program. This in turn allows the use of those resources in other areas of production and/or other products. Ultimately, a sensory program should only be evaluating product attributes that 1) cannot be instrumentally measured and 2) are critical to consumer acceptance. These modifications can only be made through the design and execution of research projects.

Sensory Coordinator's Development
The sensory coordinator's development should be in several areas: his or her own sensory knowledge, his or her sensory skills as a panelist, and his or her interaction with R&D groups and all the groups in the plant.

It is important that the sensory coordinator expands his or her knowledge in all areas of sensory evaluation. Many times, the coordinator's managers do not support this growth, since an in-plant program is considered a routine program requiring little improvements once it is in operation. However, a sensory coordinator needs to expand his or her knowledge in panel training and leadership, data analysis, and other sensory areas related to the type of evaluation and research being conducted. For that, there are a variety of short courses offered through universities, professional associations, and consulting firms that should be considered. In addition, attending professional meetings and interacting with the R&D sensory department keeps the sensory coordinator informed of the new developments in the field.

The sensory coordinator's skills as a panelist have to be expanded as well. This aspect will be covered later under the panelists' development.

The interaction with various R&D groups and all groups in the plant are also part of the coordinator's development. Maintaining an interaction with R&D fosters the coordinator's involvement in research projects that benefit the plant and the company. In addition, it allows the coordinator to get involved in R&D sensory projects and programs, such as product training and audit programs, technical projects, and so forth.

Part of the sensory professional's development is his or her interaction with all plant personnel, which allows the sensory coordinator to discuss and resolve panelists' attendance problems, gives visibility to the sensory program and its purposes, helps recruit new panelists, and fosters the interaction among supervisors, plant workers, and other plant staff.

Panelists' Development
The panelists' development should focus on product evaluation skills and their knowledge in sensory evaluation. The importance of improving the panelists' skills was discussed under panelists' motivation. Improving their product evaluation skills through training in other products not only serves as a motivational factor, but improves their technical expertise as panelists for products evaluated already, new products, and special sensory problems that may develop at the plant.

In addition, the sensory coordinator should develop a program geared to expand the panelists' sensory knowledge. This may be done by organizing presentations of basic sensory topics and of R&D sensory projects, and by circulating basic sensory technical literature.

OTHER VERSIONS

This chapter has presented the characteristics and the steps to implement the most complete version of the comprehensive descriptive method. This version requires:

- Committing to all the resources needed in this approach;
- Using it mainly for finished products with about 10 to 15 attributes evaluated; and
- Introducing one product at a time into the program.

There are, however, abbreviated versions of the comprehensive descriptive method. These are:

- A shorter, less comprehensive approach used by itself;
- A shorter version across several similar products; and
- A shorter version used in combination with other methods.

Shorter Version Used By Itself

A shorter version of the comprehensive method involves the evaluation of only a few attributes (one to five), which are the *most critical* attributes in the product. This approach works particularly well when assessing raw ingredients or very simple products (e.g., hard candy, powdered drinks). The few attributes are either chosen by management or through a research project geared to assess the importance of the attributes based on consumer acceptance (in the case of finished products) or the final product quality (in the case of raw ingredients). The process for selecting attributes is the same as that described for the complete version in previous sections. The philosophy and the evaluation procedures in this version are the same as for the complete approach. However, only the most variable and critical product attributes are chosen. The panel is trained to evaluate the intensity of only the selected attributes, using an abbreviated ballot (see Fig. 3.18A and 3.18B).

This version, because of its simplicity, requires considerably less time and resources to implement. Typically, companies that use this modified version implement it to evaluate many raw ingredients or simple products. It should be noted that a complete comprehensive descriptive approach, like the potato chip program described in this chapter, could develop into this shorter version long term. This situation is desirable and likely to occur when research has been conducted to substitute the sensory measurements by the more simple, less expensive, and less cumbersome instrumental evaluation shown to correlate with the sensory responses.

Shorter Version Across Several Similar Products

A company that produces a variety of products within the same category (e.g., a variety of salad dressings, liquid soaps with different scents) might consider implementing this modified version. In this version, all or many of the products in the group are introduced simultaneously into the program because of their similar sensory characteristics. Only a few attributes in each product type are evaluated, and there might be one or two attributes common to all products. With all products having similar characteristics and differing only in two or three sensory attributes, it is beneficial and effective to simultaneously train a panel to evaluate all products. The training first covers the common attributes and then focuses on the few attributes unique to each product. The program is effective, regardless of the company's production schedule—that is, varieties may be produced simultaneously or in a staggered manner. The panel, however, can operate on a routine basis, evaluating whatever product variety is currently being produced.

Figure 3.19 shows the evaluation forms of three different mouthwashes. Most of

A. RAW INGREDIENTS

    FRAGRANCE OF BABY POWDER

    CHARACTERISTIC   SCORE

    Total intensity of fragrance   _____

            . Sweet aromatic/vanilla   _____

            . Floral/rose   _____

    Carrier/base fragrance   _____

B. SIMPLE FINISHED PRODUCT

    BABY POWDER

    CHARACTERISTICS   SCORE

    Brightness of color   _____

    Fragrance intensity   _____

    Base odor intensity   _____

    Smoothness on skin   _____

    Residue retention   _____

FIGURE 3.18.   Evaluation form used for a modified comprehensive description method (shorter version).

the sensory characteristics are common to all varieties, except for the flavor type and color.

Shorter Version Combined with Other Methods

A shorter version of the comprehensive descriptive approach can be used in combination with other methods described in this book. This modified version

MOUTHWASH

PEPPERMINT

	0	10

Clarity

Color intensity
(blue)

Peppermint Intensity

Sweet

Cool

Burn

CINNAMON

	0	10

Clarity

Color intensity
(red)

Cinnamon intesity

Sweet

Cool

Bite/heat

Burn

CITRUS

	0	10

Clarity

Color intensity
(yellow/orange)

Orange/citrus intensity

Sweet

Sourness

Cool

Burn

FIGURE 3.19.   Evaluation form used for a modified comprehensive descriptive method (shorter version across similar products).

would be implemented when a company is interested in obtaining some attribute information in addition to other sensory measurements. Some examples are:

- Descriptive attribute evaluation after an overall quality product rating;
- Descriptive attribute evaluation after an "in/out"-of-spec assessment; and
- Descriptive attribute evaluation after an overall difference-from-control measurement.

These modifications are discussed in more detail in Chapters 4 ("Quality Rating"), 5 (" 'In/Out'-of-Spec") and 6 ("Difference from Control"), in sections discussing other versions of these methods.

4

Quality Ratings Method

ABSTRACT

This sensory program consists of assessing the quality of daily production by a panel, based on either the panelists' own perception of "quality" or established quality criteria. Samples are rated using a quality scale (e.g., very poor to excellent), and a product is rejected when the quality ratings are low. The quality score or cutoff point that determines when a product is to be considered acceptable or unacceptable is a management decision.

The main advantage of this method and the reason for its popularity is that it provides a direct measure of the product's quality and may be an appropriate selection for those companies where quality assessments are commonly understood. Another advantage of this program is the motivational factor associated with the type of product evaluation completed. Rating production's quality directly makes panelists perceive a higher degree of involvement in this program than in other programs.

This program has several disadvantages. In the absence of training, and without common quality criteria, the main limitation of this program is the high panel variability observed in the data due to differences in individual experiences and familiarity with production. When quality criteria are established and taught to panelists, the disadvantage is the cumbersome procedures that have to be followed to establish this type of program. This may translate into a long training period, frequent panel reviews, and practice sessions. Another disadvantage is that the quality measurements may be integrated and therefore may not be very actionable and useful for product documentation or guidance. For example, a "poor" peppermint quality rating in chewing gum can mean that the peppermint is low or high in intensity, has off notes, or differs in character, compared to the standard.

PROGRAM IN OPERATION

Quality ratings programs in consumer products companies vary in several ways, including the number of participants in the panel, the nature of the panel training, the type of quality criteria used in the evaluations, and the number of product characteristics assessed. The operation of two very different quality panels is described in scenarios A and B, following.

Scenario A: The panel consists of two company employees, one of whom is the QC plant manager, who evaluate production samples (up to 20 to 30 products/session) without formal and standardized evaluation procedures. Panelists rate only one quality attribute (usually "overall" quality) using a quality scale (e.g., very poor–fair–good–excellent), grade products based on their personal quality criteria, make decisions on the product disposition based on a consensus agreement, and usually do not reject products unless their scores reach the "very poor" category.

Scenario B: The panel consists of 8 to 12 panelists who are trained in the procedures to assess the quality of a given product type. During their training, panelists were taught how to rate the quality of the sensory characteristics that affect the product's quality. In addition, they learned the company's quality guidelines, which were established using the input from consumers and management. These guidelines are meant to standardize the criteria used by panelists in their evaluation. These guidelines are shown to panelists by actual products representing various quality levels. The program was designed by a sensory professional using sound methodology and adequate testing controls for the evaluation process. In routine evaluations, panelists rate "overall" quality, as well as the quality of selected sensory attributes using a balanced scale (e.g., very poor to excellent). The data are treated like interval data, and panel means are used to summarize the results of the evaluations. The results on quality are provided to management, which makes decisions on the disposition of the production batches evaluated.

These two panel cases describe the "worst" and the "best" situations encountered at plants that use the quality ratings program. Unfortunately, scenario A is more frequently encountered. The material presented in this section focuses on the approach described in scenario B, which represents a more technically sound approach to rate quality. Hopefully, companies with programs similar to those described in scenario A will consider the procedures described in this section as an alternative to improve their current, less controlled programs. The following represents a brief description of a program in operation as recommended in this section.

Table 4-1 shows the results of the quality evaluation of a paper towel production sample (B2-0115) by the panel described in scenario B. The panelists rate the "overall" quality, as well as the quality of attributes using a quality scale from 0 to

Table 4-1.   Average panel quality ratings for production sample B2-0115 of variable paper towel attributes

Attribute	Sample B2-0115 Quality Score *
Chroma	8.1
Specks	5.4
Emboss depth	9.7
Grittiness	4.1
Stiffness	3.7
Tensile Strength (D = Dry)	7.8
Paper Moistness	9.5
Overall Quality	5.8

* Rated on Scale
0 = very poor
10 = excellent

10 (0=very poor and 10=excellent). Specific attributes are rated because they were found to impact consumer acceptance. In rating this and other production samples, panelists use the quality criteria taught during their training. Quality references are reviewed periodically to reduce the variability of the panel quality ratings. Without common quality criteria, results vary widely from person to person and are dependent on each panelist's previous product experience, personal preference, knowledge of the process, and so on.

The data in Table 4-1 are provided to QC management to make decisions on the disposition of the evaluated production batch. Decisions are based on the comparison between the panel results and the quality specifications set for each of the product attributes. The last column of Table 4-2 shows the quality specifications set by management. The specifications were added to the table for the purpose of this discussion. The quality specifications are *never* included in the panelists' evaluation forms (ballots) and should not be information provided to panelists, since they may bias their ratings. The information is only known by management to compare the panel results to the product's specifications.

A product is considered unacceptable ("out-of-spec") if it falls outside the quality specifications. For example, Table 4-2 shows that production sample B2-0015 is unacceptable because the "overall" quality rating and the quality ratings of the attributes specks, grittiness, and stiffness are lower than the set minimum quality scores.

IMPLEMENTATION OF THE PROGRAM

A description of the steps followed to implement the recommended program (Scenario B) is presented below. This approach overcomes several of the limita-

Table 4-2.   Comparison of panel results with sensory quality specifications

Attribute	Average Quality Score*	Sensory Quality Specs
	(sample B2-0115)	(minimum acceptable)
Chroma	8.1	5.0
Specks	5.4	6.5
Emboss Depth	9.7	5.0
Grittiness	4.1	6.0
Stiffness	3.7	6.0
Tensile Strength (D)	7.8	4.5
Paper Moistness	9.5	7.0
Overall Qualtiy	5.8	6.0

* Rated on Scale
0 - very poor
10 - excellent

tions of existing and/or commonly encountered programs (Spencer 1977; Waltking 1982; Ke et al. 1984; Bodyfelt et al. 1988). However, the proposed approach is not the most effective and sound method for QC/sensory measurements. Despite its improvements, the quality ratings method is still cumbersome, involved, and has limitations compared to other methods presented in this book. The quality ratings method is discussed and presented in its more sound version due to the popularity of this approach in industry. Companies that use the quality ratings approach should either implement the practices in this chapter, or, preferably, consider another method in this book as an alternative to their existing QC/sensory program.

A company that selects the quality ratings method for their QC/sensory program has the following objective and has collected the following information in preliminary stages (Chapter 2):

1. Overall objective—To establish a program that measures the quality of raw ingredients or finished products, using a trained panel familiar with management's quality criteria.
2. Preliminary information and special considerations:
   a. Production samples have consistently exhibited perceivable differences (i.e., variability) over time.
   b. The differences have been identified in terms of sensory attributes and sometimes in terms of the variability (intensity) ranges of those attributes.
   c. It has been decided that documentation on product fluctuations in terms of sensory attributes is not needed. Therefore, the data generated by this program are limited to making decisions on product disposition only.
   d. Products representative of production variations can be easily obtained in any product survey, or produced otherwise for training purposes.

    e. A sensory panel is available to provide the descriptive (intensity) documentation of the samples used in the program.

    f. The resources to implement this type of program are available at both the production and research facilities (refer to "Identifying Resources" on pg. 28 in Chapter 2).

The four steps to implement this program are:

1. The establishment of sensory specifications and quality guidelines;
2. The training and maintenance of the plant panel(s);
3. The establishment of a data collection, analysis, and report system; and
4. The operation of the ongoing program.

Establishment of Sensory Specifications and Quality Guidelines

The sensory specifications for this approach are the minimum quality levels tolerated overall and for each attribute (Table 4-2).

There are two ways by which sensory specifications for this type of program are established: 1) specifications set through consumer's and management input or 2) specifications set based on management's criteria only. A description of both procedures follows. The first of these two approaches is preferred.

Specifications Set Through Input of Consumers and Management

The seven steps followed to set specifications through this approach are:

1. Assessment of needs and planning phase;
2. Collection and initial screening of a broad array of production samples;
3. Descriptive evaluation to characterize the samples;
4. Selection of samples to be tested by consumers;
5. Planning and execution of the consumer research test;
6. Data analysis to establish the relationship between the consumer and descriptive data sets; and
7. Establishment of final specifications *and* quality guidelines based on test results and management input.

The process is the same as that described for the comprehensive descriptive approach (Chapter 3), except for step 7. In this step, the sensory specifications are set in terms of quality scores, and quality guidelines are developed. The guidelines are established by management for panelists to rate quality and are demonstrated to the panelists by quality product references. Some of the procedures described in this section are relatively cumbersome. This is especially true for the establishment of quality criteria. Once step 7 is finalized, the rest of the program implementation and steps are easily completed.

The process to establish sensory specifications is illustrated for a paper product (paper towel). Only the steps that are unique to the product and this method are discussed. The reader is referred to the detailed description of the steps common to all methods that were described in the comprehensive descriptive method (Chapter 3).

Assessment of Needs and Planning Phase for the Consumer Research Study

See "Specifications Set Through the Input of Consumers and Management" on pg. 56 in Chapter 3.

Sample Collection and Initial Screening of a Broad Array of Production Samples

Following a complete survey, a total of 85 paper towel products are collected. These products are informally evaluated (screened) to select a subset of products that deviate from what is known to be "typical" of the majority of the samples collected. Twenty paper towels are chosen as representative of the production variability.

Descriptive Analysis

The descriptive data are used for three purposes:

1. The selection of the final set of products to be consumer tested (pg. 114);
2. The establishment of specifications (pgs. 115 and 116); and
3. The selection of replacement product references in later stages of the program operation.

The complete descriptive characterization is obtained for all samples (e.g., 20) collected and screened in the previous steps. Table 4-3 shows the list of descriptive terms evaluated for paper towels and the average panel results for one sample. (The scale used is a 0 to 15 intensity scale, where 0=none and 15=extreme.) The panel results obtained in the evaluation of 20 samples are summarized, and the variability ranges for each attribute are plotted in Figure 4.1. The line scales are 15 cm scales, where the distance between the left end marked "0" and the first slash mark on the line represents the lower range intensity value. Some of the attributes showing the largest variability ranges are depth of embossing, grittiness, and paper moistness.

The descriptive data for the 20 paper towel samples is analyzed for sample selection purposes.

Table 4-3.   Descriptive characterization of a paper towel production (sample # 651)

Attribute	Intensity score* (sample #651)
APPEARANCE	
Color Intensity	1.2
Chroma	12.4
Specks	1.7
Emboss Depth	6.5
DRY TACTILE	
Overall Surface Complex	11.9
Fuzzy	9.3
Gritty	2.4
Grainy	5.1
Thickness	6.3
Fullness/Body	5.7
Force to Gather	4.1
Force to Compress	5.3
Stiffness	6.5
Noise	3.1
Compression Resilience Intensity	6.7
Compression Resilience Amount	10.5
WET TACTILE	
Paper Moistness	5.2
Tensile Strength	2.3
Water Release	2.7

* Scale 0 = none 15 = extreme

Sample Selection

The analyses used for sample selection are described in "Sample Selection" on page 60 in Chapter 3. Following these steps for the paper towel examples, a total of 11 products were selected for the consumer test.

Planning and Execution of the Consumer Research Study

A consumer research study is designed to assess the impact of the production variability on consumer acceptance. In this study, consumers evaluate the 11 paper towels that span the complete production variability. Overall acceptance, acceptance to specific attributes, and attribute diagnostic or intensity data are obtained from this study. "Planning and Execution of the Consumer Research Study" on

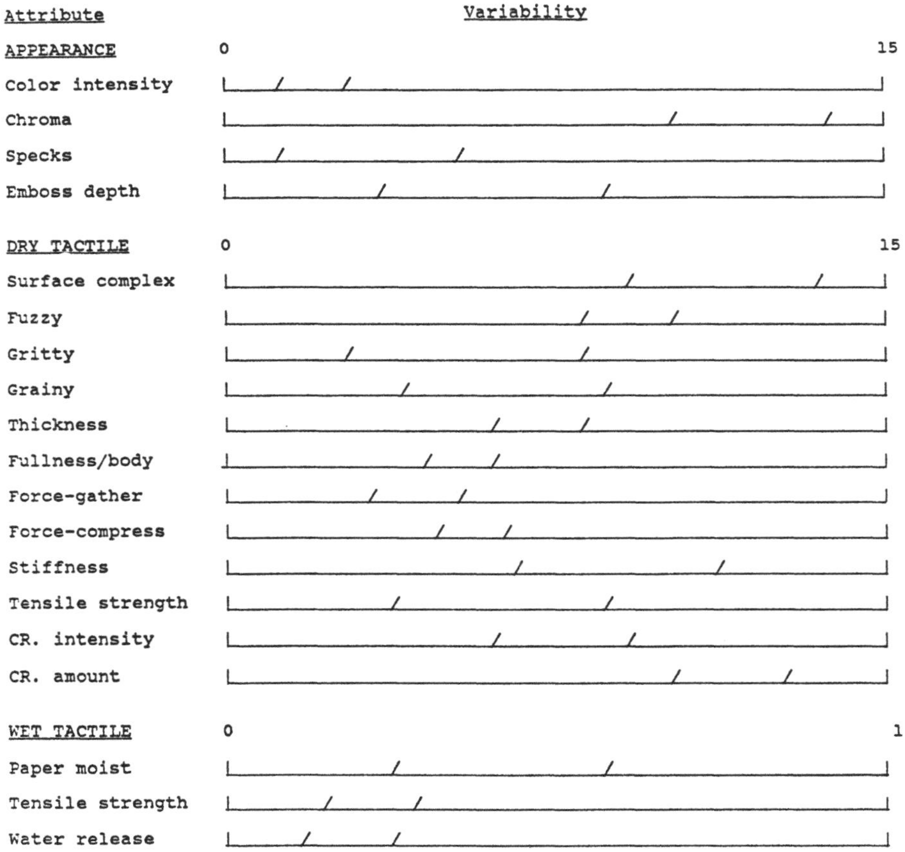

Attribute			Variability		

APPEARANCE   0   15

Color intensity └──/────/──┘

Chroma └──────────────────────────────/───────/──┘

Specks └──/────────────/──────────────────────────┘

Emboss depth └──────────────/──────────────/────────────┘

DRY TACTILE   0   15

Surface complex └────────────────────────/──────────/──┘

Fuzzy └─────────────────────/──────/────────────┘

Gritty └──────/─────────────/────────────────────┘

Grainy └──────────/─────────/────────────────────┘

Thickness └────────────────/────/───────────────────┘

Fullness/body └────────────/────/───────────────────────┘

Force-gather └────────/────/───────────────────────────┘

Force-compress └───────────/────/────────────────────────┘

Stiffness └──────────────────/──────────/───────────┘

Tensile strength └──────────/────────────/────────────────┘

CR. intensity └──────────────────/──────────/───────────┘

CR. amount └───────────────────────────/────────/──┘

WET TACTILE   0   1

Paper moist └────────────────/──────────/─────────────┘

Tensile strength └──────/────────/──────────────────────────┘

Water release └──/────────/──────────────────────────────┘

FIGURE 4.1.   Production of variability of sensory attributes of paper towels.

page 71 in Chapter 3 describes in detail the important issues to consider and the steps followed to complete the study.

Data Analysis/Data Relationships

The two data sets obtained—the descriptive data (pg. 113) and the consumer data (pg. 115)—are analyzed as explained in Chapter 3, "Data Analysis/Data Relationships" (pg. 76).

These results are plotted and prepared for the management meeting to set quality sensory specifications. Figure 4.2 shows the plots that depict the three cases of data relationships that can be found. Figure 4.2A shows the lack of an effect of attribute variability on consumer acceptance. Some of the attributes that fall into this category are gloss, fullness/body, and compression resilience intensity. Figures 4.2B and 4.2C, show linear (positive or negative) and curvilinear relationships respectively, found in the study. Grittiness showed a linear relationship and chroma showed a curvilinear relationship with acceptance.

Table 4-4 summarizes the results obtained from the data analysis. Most of the sensory attributes that had an impact on consumer acceptance showed a linear relationship with consumer acceptance.

Establishment of Sensory Specifications and Quality Guidelines
A management meeting is scheduled to review the data and establish several program parameters. The objectives of the meeting are:

• To review the relationships between consumer and descriptive data in order to identify those attributes related to consumer acceptance;
• To establish the sensory specifications in terms of quality ratings; and
• To establish quality guidelines and to select quality references.

Review of Consumer Data and Data Relationships.   The summary of the consumer test results and the data analysis is presented to management. Only the attributes that show a relationship with consumer acceptance (Fig. 4.2B and 4.2C) are used to set specifications and quality criteria. However, some companies may decide to set specifications for a few attributes that did not show a relationship with consumer acceptance, but are considered important descriptors of the product category.

If available, products from the consumer study should be presented to management during the discussion of results. The products should be examples of the lowest, highest, and one or two intermediate intensities per attribute.

Establishment of Sensory Quality Specifications.   The sensory specifications set for the comprehensive descriptive approach (Chapter 3) are "intensity" specifications, compatible with the type of data obtained (i.e., attribute intensity ratings). For example, the intensity specification for color intensity for the potato chip example is 3.5 to 6.0 (Table 3-8).

In the quality ratings method, "intensity" specifications for the critical attributes need to be set initially, followed by the establishment of "quality" specifications. A two-step process needs to be followed for two main reasons:

1. All the product information collected in prior steps (pp. 113–115) are intensity data (i.e., descriptive and consumer data).
2. Prior to this management meeting, no association exists between the product attribute intensities and the corresponding product quality.

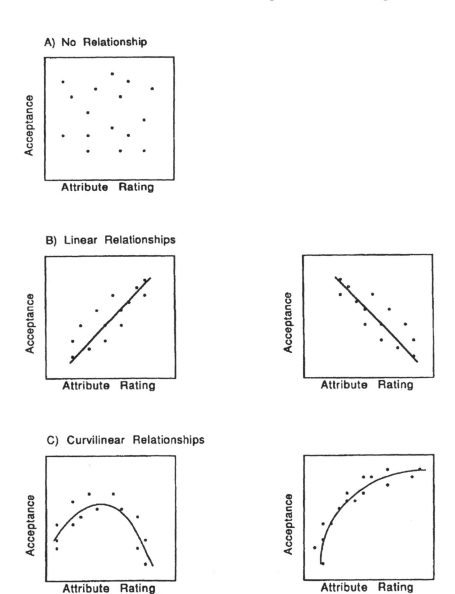

FIGURE 4.2.   Relationships between acceptance and descriptive attribute ratings found in paper towels.

Table 4-4.   **Summary of the sensory attributes related to consumer acceptance of paper towels**

Attribute	Type of Relationship*		
	Positive (linear)	Negative (linear)	Curvilinear
Chroma		x	x
Specks		x	
Emboss depth	x		
Grittiness		x	
Stiffness		x	
Tensile strength	x		
Paper moistness		x	

* Examples of relationships are displayed in Figure 4.2.

The process is described for the paper towel example and shown in Figure 4.3. For the paper towel example, it was found that both "overall" and "softness" consumer acceptance decrease with high stiffness intensities (Fig. 4.3A and 4.3B). Given an "overall" acceptance cut-off point of 6.0, the intensity sensory specification for stiffness is set as an intensity of 9.0 or lower (Fig. 4.3A).

These preliminary intensity specifications are used to set quality specifications. Figure 4.3C shows that the quality specification for stiffness, based on the intensity specs, is set as a range from 6 to 10 in quality (0 = very poor, 10 = excellent). Samples that are stiff are rated low in stiffness quality, with samples rated 6.0 or lower being considered "out-of-specification."

The quality specification for other critical attributes are established, following the same procedure. Table 4-5 lists the final quality specifications set for paper towels.

Establishment of Quality Guidelines/Selection of Quality References.
Quality guidelines are established in the management meeting to provide criteria to panelists for evaluating quality. The quality guidelines for a given product attribute specify the relationship between the quality of the attribute and the attribute intensity/level. For example, if the quality guideline for grittiness in paper towels is an inverse relationship between the quality of grittiness and grittiness intensity, then the quality of grittiness decreases as grittiness intensity increases. The quality guidelines are best demonstrated through product references that represent levels of quality. Therefore, the establishment of quality guidelines is completed through the selection of products chosen as quality references for the various critical attributes in the program.

The selection of products (i.e., quality references) to represent quality guidelines is completed in the management meeting. The products are then used for training the panel. For perishable products (e.g. foods), fresh samples representing

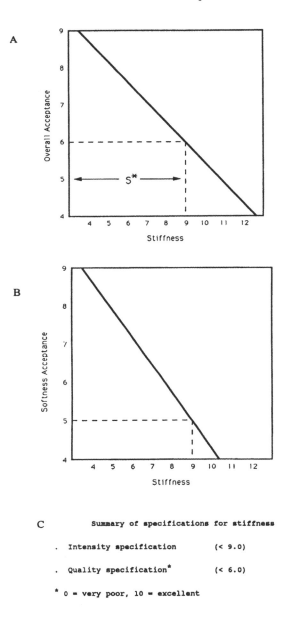

FIGURE 4.3.   Establishment of quality specifications based on intensity specifications. A) Relationship between overall acceptance and stifness intensity. [Intensity specifications set (S*) is lower than 9.0.] B) Relationship between softness acceptance and stiffness intensity. C) Summary of specifications for stiffness.

Table 4-5.   **Final sensory quality specifications for paper towels**

Attribute	Specification (minimum acceptability)
Chroma	5.0
Specks	6.5
Emboss Depth	5.0
Grittiness	6.0
Stiffness	6.0
Tensile Strength	4.5
Paper Moistness	7.0
Overall Quality	6.0

quality criteria have to be obtained at a time close to training. A small survey, a screening of samples, and the descriptive analysis of those samples have to be completed to assure that the replacement samples match the sensory characteristics of the products originally selected as quality references.

The establishment of quality guidelines is an essential step in this type of program. These guidelines are the common quality criteria taught to the panel to rate quality. With these guidelines, all panelists have common quality criteria to judge production samples. This makes the evaluation less variable and more reliable.

The steps followed in establishing quality guidelines and selecting quality references are:

- Identifying the full quality range per attribute;
- Selecting quality references for the extreme and perhaps intermediate quality points; and
- Rating each reference in "overall" quality.

Table 4-6 illustrates the process for two paper towel attributes: stiffness and tensile strength. The quality ranges chosen, based on the variability intensity ranges, are 2 to 10 for stiffness and 3 to 10 for tensile strength. Three quality references are selected for each attribute. In addition, an "overall" quality score is assigned to each product reference. The assessment of the "overall" quality scores across variable attributes is to give panelists an indication of the impact of each attribute on the total product quality. Table 4-6 shows that stiffness variability has a greater impact on overall quality of paper towels than tensile strength variability.

During this product review, it is critical to discuss the concept of "excellence" in quality control as compared to product excellence in research and development and marketing. A product may be rated "excellent" in a quality control situation

Table 4-6. Establishment of quality guidelines and selection of quality references for stiffness and tensile strength in paper towels

Stiffness

Sample ID	425			979			614
Intensity Range* (production variability)	6	7	8	9	10	11	12
Quality Range**	10			6			2

Reference Selected and Overall Quality

Sample	Stiffness Quality	Overall Quality
425	10	9
979	6	7
614	2	4

Tensile Strength

Sample ID	321	173			817
Intensity Range (production variability)	4.5	5.5	6.5	7.5	8.5
Quality Range	3	4.5			10

References Selected and Overall Quality

Sample	Tensile Strength Quality	Overall Quality
817	10	8.5
173	4.5	7.0
321	3	6.0

*Intensity Scale (0=none, 15=extreme)
**Quality Scale (0=very poor, 10=excellent)

(i.e., relative quality score compared to all production samples produced at a plant), and only "fair" in a marketing application (i.e., absolute quality score compared to all products in the market place).

An illustration of this concept is presented in Figure 4.4. Case I shows the differences in absolute sensory quality among five different commercial products (brands A through E). The distribution shows that product E has a higher absolute sensory quality compared to product B. However, from a quality control perspective using a quality ratings program (case II, Fig. 4.4) product B would obtain higher "internal" quality scores more often than product E. This is a reflection that there is less production variability in product B (range B'B*) than in product E (range E'E*).

In the management meeting, where specifications and quality guidelines are set, "excellent" samples in each attribute have to be selected (e.g., samples 425 and 817

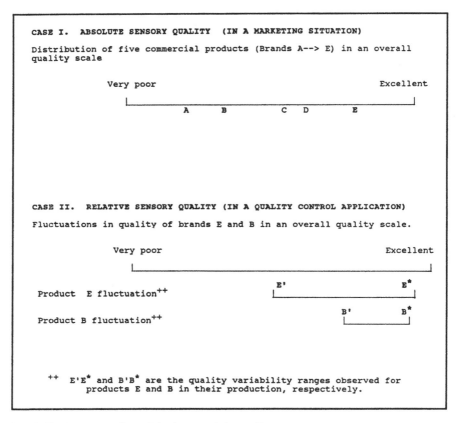

CASE I. ABSOLUTE SENSORY QUALITY (IN A MARKETING SITUATION)

Distribution of five commercial products (Brands A--> E) in an overall quality scale

Very poor   Excellent

A   B   C   D   E

CASE II. RELATIVE SENSORY QUALITY (IN A QUALITY CONTROL APPLICATION)

Fluctuations in quality of brands E and B in an overall quality scale.

Very poor   Excellent

Product E fluctuation[++]   E'   E*

Product B fluctuation[++]   B'   B*

[++] E'E* and B'B* are the quality variability ranges observed for products E and B in their production, respectively.

FIGURE 4.4.   Comparison of absolute vs. relative quality scores.

with quality scores of 10 for the attributes stiffness and tensile strength, respectively). The products represent the "best" attribute intensities that can realistically be produced.

Evaluation Form.   Based on the decision reached at the management meeting, the final evaluation form for routine assessments is developed. Figure 4.5 shows the evaluation form to be used in the paper towel program, based on the decisions made at the management meeting. Figure 4.6 shows a similar ballot for the quality evaluation of another consumer product—orange juice from concentrate. As with the paper product, the quality of each of the critical attributes (e.g., which impact on consumer acceptance) is rated, as well as the "overall" product quality.

```
                                        Name _____
                                        Product ID _____

INSTRUCTIONS

. Please evaluate the quality of the attributes listed below, using
  the scale

        0 ----------------------> 10
     very                         excellent
     poor


. Rate the quality of these attributes, using the quality guidelines
provided during training.


. Rate the OVERALL QUALITY of the products, based on your assessment of
individual characteristics.


EVALUATION                    QUALITY SCORE

   Attribute                  (0=very poor to 10=excellent)

   Chroma                     _____

   Specks                     _____

   Emboss Depth               _____

   Grittiness                 _____

   Stiffness                  _____

   Tensile Strength Dry       _____

   Paper Moistness            _____


   OVERALL QUALITY            _____
```

FIGURE 4.5.   Evaluation form used for quality ratings of paper towels.

Product Specifications Set Based on Management Criteria Only

Procedure

The previous section discussed how the consumer responses to production variability were used as guidelines to set specifications by management. Without the consumer information, management uses only their criteria for setting specifications.

To follow this approach, only the results from the product survey are required. A management meeting is scheduled where the following aspects are presented:

- The results of the product survey; and
- A subset of the production samples that demonstrate the ranges of production variability.

```
                                              Name _____
                                              Product ID _____

  INSTRUCTIONS

  . Please evaluate the quality of the attributes listed below, using
    the scale

        0 --------------------> 10
      very                      excellent
      poor


  . Rate the quality of these attributes, using the quality guidelines
  provided during training.


  . Rate the OVERALL QUALITY of the products, based on your assessment of
  individual characteristics.


  EVALUATION                    QUALITY SCORE

    Attribute                   (0=very poor to 10=excellent)

    Color Intensity             _____

    Total Orange Flavor         _____

    Fermented Orange            _____

    Sourness                    _____

    Thickness                   _____

    Pulpiness                   _____


  OVERALL QUALITY
```

FIGURE 4.6.   Evaluation form for quality ratings of orange juice from concentrate.

For the paper towel example, the results of the product survey are shown (Fig. 4.1). In addition, the products that represent the extremes of the variable attributes and some intermediate points are presented (Fig. 4.7, circles).

Each attribute is reviewed by demonstrating the products that display its variability range. Upon this review, the following program parameters are established:

- The attributes to be evaluated in the program;
- The quality ranges for those attributes;
- Quality product references; and
- Quality specifications.

(See "Establishment of Sensory Specifications and Quality Guidelines" on page

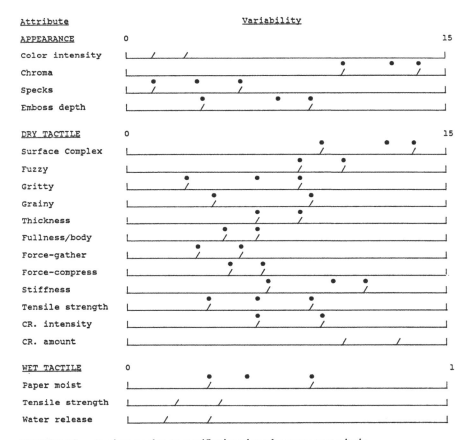

FIGURE 4.7.   Products used to set specifications through management criteria.

118 in this chapter for the considerations and steps followed to complete each of these steps.)

For the paper towel example, the attributes selected for the program are specks, grittiness, stiffness, and tensile strength. The quality ranges, references, and specifications selected for the attributes specks and grittiness are shown in Table 4-7.

Descriptive Analysis

The quality references selected to represent quality scores (Table 4-7) need to be evaluated by a descriptive panel. The references' sensory characteristics need to be documented for future steps of the program. This documentation of original quality

Table 4–7.   Example of program parameters established by management for specks and grittiness in paper towels

Parameters	Values/Samples	Scale*
Specks		
Intensity range	1–5.0	0–15
Quality range	10–5	0–10
Quality references	Sample 615 (quality = 10)	
	Sample 313 (quality = 4)	
Quality specification (minimum acceptable)	6.5	0–10
Grittiness		
Intensity range	2.5–8.0	0–15
Quality range	10–2	0–10
Quality references	Sample 212 (quality =10)	
	Sample 813 (quality = 2)	
Quality specification (minimum acceptable)	6.0	0–10

*0–15: Intensity ranges (0 = none, 15 = extreme)
 0–10: Quality scales (0 = very poor, 10 = excellent)

references (data base I) is compared to the characterization of replacement production samples (data base II) to assure that the new references match the original products. A continuous supply of references is needed for panel recalibration and periodic reviews.

Training and Operation of Plant Panels

As with any in-plant program, the panel can be formed by plant employees or local residents. The advantages and disadvantages of these two options are covered in Chapter 3.

Training Steps

A prescreening and a screening phase are conducted to identify and select the panel members ("Prescreening" and "Screening" sections on page 85 in Chapter 3.) The training is scheduled to train 20 to 30 selected participants. Overall, the training program for the quality rating approach consists of a basic sensory program and a product training phase. In the product training phase, panelists are trained to evaluate the quality of those characteristics that either (1) were shown to impact consumer acceptance or (2) were chosen by management as being critical attributes. In addition, common quality criteria are established among the panel members through the demonstration of quality references selected in previous steps. Depending on the product category and the number of attributes, the training program requires six training sessions of 2 to 5 hours each, for a total of 12 to 30 hours. Additional sessions are scheduled thereafter for practice.

For the paper towel example, the attributes taught to panelists are:

- Chroma
- Specks
- Depth of embossing
- Grittiness
- Stiffness
- Tensile strength (dry)
- Paper moistness

In addition, guidelines are given to rate the "overall" quality of daily production samples.

Sample Selection for Training
The samples used for training are the quality references chosen in the management meeting (pg. 118) to represent attribute quality levels. These product references 1) are available from the original product survey and management meeting or 2) need to be obtained through a new survey.

For the paper towel example the training samples are available, since no product changes occur during storage and a large enough product quantity is stored initially to complete several program steps. Figure 4.8 shows the samples (quality references) used in the training program. The products are the samples chosen to represent extreme and intermediate quality scores for the various attributes to be evaluated.

For perishable products, new references may be needed for the training phase, due to the product changes upon storage. A new product survey is scheduled for this purpose, or pilot plant samples are produced. The descriptive analysis data base (pgs. 113 or 125) is used to find new production samples, whose sensory characteristics match those of the quality product references.

Training Program
The training program consists of a basic sensory evaluation phase and a product training phase. The objective of the basic sensory program is to familiarize

ATTRIBUTE   QUALITY SCORE

	Very poor										Excellent
Chroma	0	1	2	3	4	**5**	**6**	7	8	9	**10**
Specks	0	1	**2**	3	4	5	6	**7**	8	9	**10**
Emboss Depth	0	1	2	**3**	4	**5**	6	7	8	9	**10**
Grittiness	0	1	2	**3**	4	5	**6**	7	8	9	**10**
Stiffness	0	1	**2**	3	4	5	**6**	7	8	9	**10**
Tensile Strength (D)	0	1	2	**3**	4	**5**	6	7	8	9	**10**
Paper Moistness	0	1	2	3	4	**5**	6	7	8	9	**10**

FIGURE 4.8.   Example of program parameters established by management for specks and grittiness in paper towels.

panelists with the applications of sensory evaluation, its importance within the company and in the measurement of quality, and the basic concepts of physiology that play a role in the evaluation of the company's products characteristics (e.g., appearance, flavor, skinfeel).

The product training phase for this approach has two components:

- Training on the detection and description of sensory characteristics (descriptive component); and
- Training on quality ratings of the above sensory characteristics (quality component).

Panelists are usually trained in the attribute definitions and evaluation procedures first (descriptive component), followed by the quality ratings component.

In the paper towel example, the definitions and evaluation procedures of the critical attributes are taught first. Table 4-8 shows the evaluation procedures taught to panelists for two of the critical paper towel attributes: stiffness and tensile strength. Products that differ in these attributes are presented during training. Although the differences between products are discussed, the training does not focus on intensity ratings, since attribute intensity measurements are not collected from the panel.

Once the panel has learned the sensory characteristics, they are trained to grade the quality of the characteristics (quality program). For that purpose, the quality product references representing extreme and intermediate quality scores are presented (Fig. 4.8).

Table 4-8.   Example of definitions and evaluation procedures of some paper towel attributes (descriptive training)

Stiffness

Definition:

Amount of pointed ridged or crooked edges, not rounded/pliable
                    (pliable/flexible——->stiff)

Procedure:

Lay towel on flat surface. Gather towel with fingers and manipulate gently without completely closing hand.

Tensile Strength

Definition:

The force required to break/tear paper
                    (no force——->high force)

Procedure:

Grasp opposite edges of towel in hands and pull the towel until it breaks.

The above sequential type of training program is unique to this QC/sensory method and makes the design and completion of the training more cumbersome and difficult than other training programs presented in this book. The product evaluation is straight forward in the comprehensive descriptive and difference-from-control methods, because only the quantification of attribute or difference intensities is taught. However, product quality evaluation as discussed in this chapter involves teaching descriptive and quality concepts in two stages.

The approach for teaching quality concepts and rating procedures is illustrated for two paper towel attributes in Tables 4-9 and 4-10. Table 4-10 shows the most complex case and demonstrates the difficulty of this method.

Table 4-9 demonstrates the simple case to evaluate attribute quality. The quality grading of attributes such as stiffness varies linearly and negatively with intensity. For stiffness, *low* intensity translates into *high* quality, and *high* intensity translates into *low* quality. This case also applies to the evaluation of fermented orange (Fig. 4.6), where the quality varies linearly and negatively with intensity: *Low* intensity of fermented orange translates into *high* quality, while *high* intensity of fermented orange translates into *low* quality.

Table 4-10 is the complex training example. For chroma in the paper towel example, the low quality score applies to *both* low *and* high attribute levels. Other examples that fall into this category are: color intensity in cookies/crackers, salt and sweetness intensities in a variety of food products, fragrance and thickness intensities in scented household products, and so on.

Table 4-9.   Approach followed to teach stiffness quality

Step 1—Definition and evaluation procedure (Table 4-8)
- Lay towel on flat surface. Gather with fingers and manipulate gently without completely closing hand, evaluate for:
  - Amount of pointed, ridged, or cracked edges, not rounded/pliable.

Step 2—Familiarization with low and high *intensities/levels* of stiffness

Sample C	Sample F
low stiffness	high stiffness
(pliable/flexible)	(stiff)

Step 3—Learning of *quality* rating through the use of product quality references

Sample C	Sample F
low stiffness	high stiffness
high quality	low quality
stiffness quality score = 10	stiffness quality score = 6

Table 4-10.   Approach followed to teach chroma quality

Step 1—Definition and evaluation procedure
- Lay towel on flat surface against dark background, visually evaluate for:
  - The chroma or purity of the color, ranging from dull/muddied to pure/bright color.

Step 2—Familiarization with low, medium, and high *intensities/levels* of chroma

Sample A	Sample S	Sample M
low chroma	medium chroma	high chroma
(dull)	(medium brightness)	(high brightness)

Step 3—Learning of *quality* ratings through the use of product quality references

Sample A	Sample S	Sample M
low chroma	medium chroma	high chroma
low quality	high quality	low quality
chroma quality score = 5.0	chroma quality score = 9.0	chroma quality score = 6.0

To train the panel to rate quality of attributes falling in Table 4-10 requires the use of:

- More quality product references (low, medium, and high intensities); and
- Two product references identified with the *same* quality levels that may be considerably different in sensory characteristics (for chroma in paper towels, samples A and M have similar quality ratings but different chroma levels—that is, sample A has low chroma, Sample M has high chroma).

The process described in Tables 4-9 and 4-10 illustrates the difficulties involved in this method. Therefore, companies that want to implement this approach need to understand this complex process before choosing this method.

After each paper towel attribute is learned, test production samples are presented for evaluation and the results are discussed. The test samples are coded with three digit codes, and panelists are asked to rate the quality of the attribute they just learned. Initially, the samples representing the largest difference in an attribute are presented as test samples. For example, for depth of embossing, the most different samples available representing the lowest (3.0) and highest (10.0) quality intensity in this attribute are chosen for the first panel exercise. After panelists have rated the two samples, the results are compared to their known depth of embossing quality intensities and are discussed with the panel. In the two to five training sessions, all critical attributes are reviewed and their quality evaluation exercises completed.

Thereafter, practice sessions are conducted to reinforce the concepts learned and to give the panelists additional practice in product evaluation. In each practice session, panelists are asked to evaluate selected production samples, presented as unknowns. Records of the evaluations are kept and used for panel feedback. Production samples identified in the survey and quality references are used for the practice sessions. The known quality ratings of the reference samples are used to judge panel performance. Both individual panelist ratings and the panel mean ratings should be compared to the known value of the reference sample to identify where, if any, problem areas exist.

Once the practice period is completed and problem areas have been resolved with the panel, the ongoing operation of the program is started. This means that the panel is qualified to rate the quality of actual production samples. Furthermore, the panel's data can then be handled with other QC data for product decisions.

Panel Maintenance

The quality ratings QC/sensory program also requires the establishment of a maintenance program, which addresses psychological and physiological considerations. Chapter 3 summarizes some of the activities involved in the maintenance program. Among the psychological aspects to be addressed are recognition and appreciation of the panelists' participation, the planning of special group activities, rewards, panel performance feedback, and panel review sessions.

The physiological factors described in Chapter 3 are also applicable in the quality ratings program. These are:

- The scheduling of review sessions. These sessions, scheduled approximately once a month, allow the review of technical and other program parameters. Among the technical aspects reviewed are attribute definitions, evaluation procedures, and quality guidelines. Other aspects of the program that need to be addressed are panel scheduling, conditions of the panel room, sample preparation, and so forth.

- The review of quality criteria. The review of quality references in this program is critical. Panelists need to be exposed to the products representing the various quality levels to assure adequate performance. This review can be done by presenting a series of products that represent the low and high quality levels for each evaluated attribute (as it was done in the training) or by presenting one product reference with all its quality scores identified. For some nonperishable products, like paper towels, large quantities of products representative of different quality levels should be obtained initially and stored under controlled conditions for later use. For most products, however, replacement quality references need to be produced/identified on a regular basis. To properly administer the acquisition (through surveys or production of pilot plant samples) and use of quality references, the sensory coordinator needs to use the descriptive data base gathered at initial stages of the program (pg. 113 or 125) and a descriptive panel that can characterize the sensory properties of the replacement quality references. The review of quality references allows panelists to recall the criteria to rate the quality of daily production. This practice also decreases within panelist variability.

- The evaluation of blind references. The introduction of blind quality references in the product sets is a practice by which the panelist performance is monitored. The results obtained for these products are compared to the documented quality ratings for that product. Individual panelist performance is monitored using control charts (see Appendix 5) and summary statistics, as shown in Table 4-11. Examination of Table 4-11 reveals that Panelist C rates the reference samples with consistently higher quality ratings than the other panelists. Further examination reveals that Panelist G is consistently more variable (i.e., larger standard deviations) than the other panelists. The observations suggest that retraining is needed for both panelists.

The Establishment of Data Collection, Analysis, and Report System

The data collection, data analysis, and report system for the quality ratings method are the same as are used in the comprehensive descriptive method. The rating scales (attribute intensities versus quality ratings) are different in nature, but are treated the same statistically.

Data Collection
Samples to be evaluated by the plant panel are collected at the same time as the samples that will undergo other QC testing. The frequency of sampling is determined using the same QC criteria that apply to all other tests.

Table 4-11.   Panelist performance statistics on blind reference samples

Panelist	Chroma		Specks		Emboss Depth		Quality Rating Scales				Tensile Strength		Paper Moistness	
							Grittiness		Stiffness					
	$\overline{X}$	s	$\overline{X}$	s	$\overline{X}$	s	$\overline{X}$	s	$\overline{X}$	s	$\overline{X}$	s	$\overline{X}$	s
A	7.2	0.8	7.0	0.7	6.9	0.6	7.1	0.8	7.3	0.7	6.8	0.7	7.2	0.8
B	6.9	0.7	7.1	0.6	7.2	0.7	7.1	0.8	7.2	0.6	7.1	0.8	7.1	0.8
C	8.2	0.6	8.3	0.7	8.2	0.8	8.1	0.9	8.4	0.8	8.2	0.7	8.4	0.9
D	6.8	0.8	6.9	0.8	7.3	1.0	7.0	0.6	6.9	0.9	6.9	0.9	7.2	1.0
E	7.1	0.8	6.7	0.9	6.8	1.0	6.9	0.6	7.0	1.0	7.2	1.0	7.3	1.1
F	7.0	0.9	7.3	0.6	6.7	0.9	6.8	0.7	6.9	1.0	7.2	0.9	7.0	0.9
G	7.0	1.8	6.9	2.3	7.1	1.7	7.0	2.2	7.1	2.4	7.0	1.9	7.1	2.3
H	6.8	0.6	7.2	1.0	6.9	0.7	7.0	0.9	6.8	0.8	6.9	0.8	7.5	0.8
I	7.1	0.7	6.8	0.8	6.9	0.8	6.9	1.0	6.8	0.8	6.8	0.7	7.3	1.0
J	6.9	0.7	7.0	0.7	7.1	0.8	6.9	0.7	7.0	1.0	7.1	0.7	7.1	0.8

The number of samples submitted to the panel per session affects the sample presentation design used by the sensory coordinator. If the number of samples per session is small, then a complete block presentation design is used. If, however, the number of samples per session exceeds that which can be evaluated before sensory fatigue sets in, a balanced incomplete block presentation design is used (see Appendix 3).

For the paper towel example, the sensory coordinator knows that four samples of paper towels are collected during each shift at approximately 2, 4, 6, and 8 hours. A shift's production is considered to be a single lot of product. The quality ratings panel of the following shift evaluates the previous shift's production. The four samples are evaluated using a complete block design with the order of presentation of the four samples being randomized separately for each of the ten panelists.

Data Analysis

For each quality rating scale, the individual panelist's scores are recorded for each sample, as shown in Table 4-12 for the chroma quality scale. The panel's average quality ratings, which appear in the last row of Table 4-12, are the raw QC data associated with each sample. For complete block presentation designs, as used for the paper towels, arithmetic means are used. For incomplete block presentation designs, adjusted (or least square) means must be computed (see Appendix 3).

The panel's average quality ratings should be recorded in a format consistent with that used for other QC tests. Specifically, if multiple samples are collected from each lot, as in the paper towel example, the lot average and range (i.e., the average and range of the four panel means) are computed and recorded (see Table 4-13). If only a single sample per shift is collected, the panel mean is the shift mean rating; no measure of within-lot variability is available.

Report System

The sensory coordinator should highlight any samples that fall out of specification, using the company's standard procedures for such occurrences. For example, in Table 4-13, the sensory coordinator has highlighted that the current lot of paper towels (#27B-13) is below the minimum acceptable quality specification for paper moistness (i.e., 7.0 from Table 4-5). The sensory coordinator forwards the information to management, who decide on the final disposition of the lot.

If an SPC program is in effect, the control charts for each of the quality ratings should also be updated and forwarded to appropriate personnel. If multiple samples per lot are collected, $\overline{X}$-charts and R-charts are commonly used to track production (see Appendix 5). If only a single sample is collected per lot, then I-charts are used in place if $\overline{X}$-charts. No analogue for R-charts is available when only one sample per lot is collected.

Four production samples of paper towels are collected every shift (i.e., lot), so

Table 4-12.   Panelists' quality ratings of the chroma of four production samples of paper towels

Lot: 27B-13
Scale: Chroma

Panelist	Sample Code Time (Hour)	714 2	312 4	294 6	563 8
A		7	8	7	8
B		7	6	6	7
C		9	10	8	9
D		8	7	6	8
E		6	6	5	7
F		5	8	7	6
G		6	8	6	7
H		8	8	6	7
I		8	7	7	6
J		6	7	5	6
Panel Mean		7.0	7.5	6.3	7.1

Table 4-13.   Panel averages and ranges of quality ratings for a batch of paper towels

Lot: 27B-13

Quality Scale	Mean	Range
Chroma	7.0	1.2
Specks	7.2	0.8
Emboss Depth	6.9	1.1
Grittiness	8.1	1.3
Stiffness	7.4	0.9
Tensile Strength	6.4	0.8
Paper Moistness	6.7*	0.8

*Lot is below minimum quality spec.

an $\overline{X}$-chart and R-chart are maintained for each of the quality rating scales used by the panel. Figure 4.9 illustrates the results for the chroma quality scale. Although each of the most recent lots are within specification (i.e., $\overline{X} > 5.0$) and, in fact, within the control limits for the process, the most recent lot is the ninth lot in a row to fall below the process mean value. Based on the criteria for control charts presented in Appendix 5, this occurrence signals that the process is out of control and should be examined to determine what is causing the shift to lower chroma quality ratings.

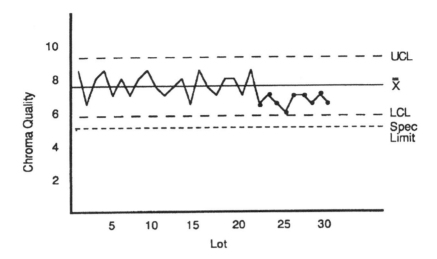

FIGURE 4.9.   X-chart of the chroma quality ratings of paper towels showing a trend toward lower quality ratings over the last nine product lots.

Characteristics of the Ongoing Program

The activities of the program in operation are grouped by daily/routine and long-term activities and are listed below (see Chapter 3 for a detailed discussion).

Daily Routine Program Activities

The daily routine program activities include:

1. Scheduling of panel sessions;
2. Scheduling of panelists' participation;
3. Sample and reference preparation;
4. Administration of test;
5. Data analysis, report, and interpretation with other QC data;
6. Scheduling of special sessions; and
7. Panel maintenance.

Long-Term Program Activities

The long-term program activities include:

1. Program growth:
   a. Introduction of new products;
   b. Introduction of other points of evaluation in the process;

c. Design and execution of research projects;

d. Improvement of sensory methods (modification of quality ratings method or introduction of new methods, such as those presented in Chapter 3, 5, and 6).

2. Sensory coordinator's development; and

3. Panelists' development.

Chapter 3 gives a description of all these activities, procedures, requirements, benefits, and special considerations.

OTHER VERSIONS

There are other quality program versions that can be developed and implemented at the plant level. The program described in detail in this chapter dealt with a program that provides information on the "overall" quality of the product and the quality of selected attributes. To establish such a program with the use of management quality guidelines is a cumbersome and expensive process. In addition, these quality results are not actionable and can fail to provide information that can be used for problem solving.

Two other versions of a quality ratings program are given below. Although only limited product quality information is obtained, these modified versions solve some of the limitations of the complete program described in this chapter.

Shorter Version: Overall Product Quality Rating

Some companies may be interested in only obtaining an assessment of the overall product quality, without requiring attribute quality ratings. Companies interested in this approach use this evaluation as a very general and quick final check of finished products only. This means that these companies rely almost exclusively on the instrumental procedures and consider the sensory assessment only as a check required to detect gross problems. A company with this interest does not need the detailed attribute information and/or is not willing to commit all the resources needed to obtain such information.

The establishment of such a program is considerably simplified if only the "overall" product quality is to be evaluated. Despite its simplicity, it is recommended that quality references also be used for the panel training. The sensory coordinator must recognize that low quality can result from a variety of perceivable characteristics. Products representing low- and high-quality levels for a variety of characteristics should be selected and used in the training.

The panelists may be trained to evaluate the "overall" product quality exclusively or asked to describe the defects in a qualitative way. Figure 4.10 shows a ballot for the quality evaluation of vanilla cookies, in which panelists rate "overall" quality and provide descriptors of any defects found.

Overall Product Quality and Attribute Descriptive Evaluation

This version combines the philosophies of the methods described in Chapter 3 (comprehensive descriptive) and the quality ratings method. This program is meant to take the best aspects and advantages of both programs:

- The assessment of the "overall" product quality. Companies obtain a quality rating, which is a measurement that may be well understood among managers and in production and research facilities.
- The collection of descriptive information for key attributes that are known to affect the overall quality and acceptance of the product. The descriptive information is actionable and has the advantages of attribute intensity data described in the comprehensive descriptive method (Chapter 3).

Although the establishment of this modified program may still require most of the resources of a full program, training in this version is less cumbersome than the process followed in the complete quality ratings program. Since the attribute information is collected as straight intensity data (descriptive data), there is no need to establish, use, and maintain quality criteria and references for attributes. Most importantly, this alternative generates descriptive data that can provide an explanation of why the product is not rated as "excellent."

Intensity references for attributes need to be selected and used as described in Chapter 3. However, this is still less cumbersome than the selection and documen-

```
.  Please rate the overall quality of this cookie.

      0   1   2   3   4   5   6   7   8   9   10
      very                                     excellent
      poor


.  Indicate the defects found, if any.

   APPEARANCE           FLAVOR              TEXTURE

   _____      _____     _____

   _____      _____     _____

   _____      _____     _____

   _____      _____     _____
```

FIGURE 4.10.   Evaluation form used for the overall quality evaluation of vanilla cookies (modified quality ratings method).

tation of quality attribute references. Because of these advantages, the authors consider this modified version to be the best option within a quality rating program.

Figure 4.11 shows the evaluation form for antiperspirant sticks, using this quality/intensity ratings modified version.

OVERALL PRODUCT QUALITY AND ATTRIBUTE DESCRIPTIVE

. Please evaluate the intensity of the following product attributes using a scale

     0   1   2   3   4   5   6   7   8   9   10
   none   extreme

SCORE

Gloss   _____

Force to spread   _____

Amount of residue (immediate)   _____

Occlusion (5 min.)   _____

Amount of residue (5 min.)   _____

Other

_____   _____

_____   _____

_____   _____

Based on the above product characterization and the quality guidelines for antiperspirant sticks, rate the overall quality of the product using the following scale.

     0   1   2   3   4   5   6   7   8   9   10
  very poor   excellent

QUALITY RATING

OVERALL QUALITY   _____

FIGURE 4.11.   Evaluation form used for the overall quality and attribute evaluation of antiperspirant sticks (modified quality rating method).

5

"In/Out" Method

ABSTRACT

In this program, daily production is evaluated by a trained panel as being either within ("in") or outside ("out") the sensory specifications (or the concept that represents "normal" production). The result of this panel is the percentage of the panelists who assess the product to be "in" specification. This method is mainly used to identify and reject products that show gross deviations from "in/typical" production, such as off-notes. It is a recommended method for the evaluation of raw materials, relatively simple finished products, or more complex finished products with very few variable sensory dimensions. In this program, panelists have a more direct participation in the decision-making process of product disposition than in other methods. Ultimately, management makes the decision on product disposition by setting an action standard around the panelists' results.

Currently, the "in/out" method may be the most popular method for QC/sensory evaluations at the plant level. Its advantages are the simplicity of the assessments, the short training and evaluation periods required, and the direct use of panel results. Comparable to the quality ratings program, panelists are highly motivated to participate in the program because of their greater involvement in the decision-making process.

The main disadvantage of the program is its inability to provide descriptive information and therefore its lack of direction and actionability to fix problems. Consequently, this method is a "decision-making" tool, rather than a source of product information. Other disadvantages include the use of panelists as judges of quality and product disposition and not of sensory characteristics, the need for a relatively large panel pool for data analysis and interpretation, and the inability to relate these data to other data, such as instrumental measurements. In addition, because of the pressure that panelists experience in making decisions regarding product disposition, the program may be subject to failure unless carefully maintained.

PROGRAM IN OPERATION

Comparable to the quality ratings program, there are two scenarios that describe the two types of "in/out" programs currently in operation in the consumer products industry.

Scenario A: The panel consists of a small group of company employees (4 to 5), mainly from the management team. The panel evaluates a large amount of production samples (up to 20 to 30) per session without standardized and controlled protocols. Each product is discussed to determine if it is to be considered "in" or "out" of specifications. In this program no defined specifications or guidelines for product evaluation exist, and no training or product orientation was held. As a result, each panelist makes decisions based on his or her individual experience and familiarity with production, or based on the highest ranking person on the panel. In this type of scenario, terms[1] such as "acceptable/unacceptable," "good/bad," "typical/atypical" might be used by the evaluators.

Another version of scenario A might be when only *one* judge evaluates products to decide if they are acceptable or unacceptable.

Scenario B: The panel consists of 25 or more panelists who evaluate products following standardized and controlled procedures and protocols (e.g., amount of products evaluated in a session, product presentation). Panelists individually evaluate production samples and assess if each product is "in" or "out" of specifications. The criteria used to make such decisions are standard guidelines established by management and taught to all panelists during training. In addition, panelists were familiarized with the characteristics that define "in-spec" products and the characteristics that define "out-of-spec" products. Therefore, all panelists work with the same criteria in the assessment of production samples. Panelists might also indicate the reason(s) why a product was assessed to be "out-of-spec." This qualitative information helps management define the condition of the "out-of-spec" product. Data are summarized as percentages of "in-spec" observations and are analyzed statistically.

These two examples represent the "worst" and the "best" scenarios found in production facilities, respectively. The situation described in scenario A represents the most frequently encountered program. This program is used by companies that want an in-plant sensory program, but lack either the resources (i.e., panelists, panelists' time, and management support) or the background in methodology to implement a more sound program. However, because of the simplicity and the very short time required to implement scenario A, companies often choose this form of

---

[1]These terms are inappropriate for analytical product evaluation. No reference to these terms (acceptable/unacceptable, good/bad, typical/atypical, etc.) is made in this book, unless it refers to the type of program described in scenario A above.

the "in/out" method. The lack of common product rejection guidelines among panelists to judge products is a shortcoming of the scenario A program. Without clear guidelines, panelists judge products based on their own criteria and preferences. This situation leads to highly variable and subjective data. Therefore, this chapter focuses on establishing defined and standard decision-making criteria for an "in/out" program (scenario B). Companies using the simple approach of scenario A can hopefully implement some of these recommendations to improve their in-plant sensory program.

The program described in this chapter has the characteristics described in scenario B above. This program has distinct differences compared to other programs described in this book (see Table 5-1). These differences make the "in/out" program unique in its implementation and operation.

A panel that operates in a situation like the one described in scenario B uses an evaluation form such as the one shown in Figure 5.1. Products are evaluated by each panelist who indicates if the product is "in" or "out" of specification based on criteria set by management.

Figure 5.2 shows another version of a ballot for this program. This version includes a comment section, where panelists indicate the reasons why that product was considered "out-of-spec." This information is qualitative in nature. Unless panelists are well trained and use standardized terms to describe deviations, this information does not provide insight into the reasons of why a product was "out-of-spec."

In their evaluations, panelists use the "in/out" product guidelines taught during training. The "in" guidelines or concept represent the characteristics of "typical/normal" production and tolerable variations. The "out" guidelines or concept represent the characteristics and intensities of products considered unacceptable or

Table 5-1.   Summary of main differences between the "in/out" method and other programs (comprehensive, quality, and difference from control)

Characteristic	Other Programs*	"In/Out" Program
Specifications known and operated by	QC or plant management	Panelists
Product decisions made by	QC or plant management	Panelists
Type of data	Ratings (Intensity, quality, or difference)	Counts
Measurement of product variability	Yes	No

* Comprehensive descriptive, quality, and difference from control methods, described in Chapters 3, 4, and 6 respectively.

■ INSTRUCTIONS

. Please evaluate the products presented to you and indicate in the
space provided if they are "in" or "out" of specification. Use the
"in/out" product guidelines taught during training to make your
decision.

Product	In	Out
_____	[ ]	[ ]
_____	[ ]	[ ]
_____	[ ]	[ ]
_____	[ ]	[ ]
_____	[ ]	[ ]
_____	[ ]	[ ]

FIGURE 5.1.   Example of an evaluation form used for the "in/out" method.

"out-of-specification." The "in" and "out" guidelines are provided by management with or without consumer input. These guidelines are demonstrated to the panelists during training, using product references.

Table 5-2 shows the results of six production batches evaluated by this method. The panel results show that two of the batches are "out-of-specification." The results are reported to management, which uses this information for the final decision on the product disposition.

The "in/out of spec" method is suitable and recommended for special situations and for certain products. This method is very useful to detect gross sensory problems or extreme fluctuations arising during production in a very efficient manner. In addition, it is most suitable for the assessment of the products/systems described below.

1. Raw materials with one attribute variation—Simple raw materials with one or two distinctly different attributes that determine the product's acceptance/rejection are suitable for "in/out" judgements. Examples: off-notes in any food material or fragrance, color or chroma of a dye, main character of a product (e.g., main flavor characters of incoming flavors, total orange impact in orange juice, total sweetness impact in syrups).
2. Raw materials or finished products with attribute variations that are not related—Simple systems such as fruit juices, marshmallows, syrups, and fabrics fall into this category. The product's attributes vary without interaction and are perceived as distinctly different attributes. For example, the variable and independent attributes of marshmallows that would determine "out-of-spec"

■ INSTRUCTIONS

. Please evaluate the products presented to you and indicate in the
   space provided if they are "in" or "out" of specification. Use the
   "in/out" product guidelines taught during training to make your
   decision.

. Use the space provided to indicate why you considered a product to be
   "out-of-specification."

	In	Out
Product _____	[ ]	[ ]

   Reasons if "out"

   _____

   _____

   _____

	In	Out
Product _____	[ ]	[ ]

   Reasons if "out"

   _____

   _____

   _____

	In	Out
Product _____	[ ]	[ ]

   Reasons if "out"

   _____

   _____

   _____

FIGURE 5.2.   Example of evaluation form used for the "in/out" method (quantitative and qualitative).

Table 5.2.   QC/sensory results on several production batches using the "in/out" method

Batch	Percent of Panel Rating the Batch "IN"
A3-0016	65.6
A3-0018	59.8
A3-0020	58.4
A3-0024	42.6[*]
A3-0026	44.8[*]
A3-0028	74.3

*Out-of-specification (Percent "IN" < 50%)

product could be vanillin intensity, size, and firmness. These attributes do not interact with one another, and, therefore, the only reference materials needed for training are products that display the levels of vanillin, firmness, and size, at which the product would be considered "out of spec."

3. Complex raw materials or finished products with many slight variations, resulting in a major negative attribute—This case includes products where the variation of many attributes results in a new perceived attribute that is considered negative, such as a cooked food product that develops a "scorched" or "burnt" note with excessive heat treatment.

4. Complex raw materials or finished products with large within variability—Examples include products that show a large batch-to-batch, container-to-container, or package-to-package variation (e.g., canned soups, frozen dinners, meat products). Attribute differences are perceived from package to package, but they are not considered negative until these differences are extreme. In their training, panelists are shown both a) variable products that are considered "in-spec" and b) variable products with more extreme differences considered "out of spec."

IMPLEMENTATION OF THE PROGRAM

The type of program described in this section has the characteristics of scenario B described above, where attention to three critical parameters of the program are given: establishment of "in/out" evaluation criteria, completion of a sound training program, and standardization of product preparation and serving protocols. This program is considered a sound method for evaluating production samples at the plant level when no detailed product information is required.

A company that selects the "in/out of spec" method for inspection of regular production or ingredients has the following objective and has collected the following information in preliminary stages (Chapter 2):

1. Overall objective—To establish a program in which raw ingredients or finished products are accepted or rejected by a trained panel based on how each product fits the concept or space of "in" or acceptable product.

2. Information gathered and special considerations:
   a. Production samples or raw ingredients have consistently exhibited perceivable differences (i.e., variability over time).
   b. The differences have been identified in terms of product attributes and perhaps in terms of the variability (intensity) ranges of those attributes.
   c. The nature of the product deviations has been assessed (see section on "Program in Operation" on pg.143) and the "in/out of spec" program is selected as the best alternative.

d. Documentation of product fluctuations is not needed. Therefore, this program is to be used for decisions on product disposition only.
e. Products representing production variations can be easily obtained in surveys or can be produced otherwise for training purposes.
f. A descriptive panel is available to provide the descriptive (intensity) documentation of the samples used for training or calibration.
g. The resources to implement this type of program are available at both the production and research facility (refer to Chapter 2 "Identifying Resources" on pg. 28 in Chapter 2).

The four steps to implement this program are:

1. The establishment of "in/out" specifications and selection of product references;
2. The training and maintenance of the plant panel(s);
3. The establishment of the data collection, analysis, and report system; and
4. The operation of the ongoing program.

Establishment of "In/Out" Specifications and Selection of Product References

As described earlier, the specifications of this program have different characteristics and uses, compared to those of other programs. Specifically, the "in/out" specifications:

- Represent the boundaries that separate the sensory space or concept of "in" product from the space/concept of "out" product;
- Are used by panelists, not management;
- Are taught to panelists through product references *without* their detailed descriptive characterization; and
- Require descriptive characterization only for documentation purposes in order to obtain new product references or interpret "out-of-spec" results.

Other chapters in this book have presented two options by which specifications are set: 1) through consumers and management input or 2) through management input only. Also, in other chapters, the first option (with input from consumers and management) was recommended.

Although either of the two procedures can be followed to establish specifications for the "in/out" method and a brief summary of both procedures is presented below, the option to set specifications by management only is usually chosen by companies that implement this program. The characteristics of this program, specifically its simplicity and limitations of the data obtained, do not justify the expensive, time-consuming, and cumbersome procedures necessary for the consumer research.

Specifications Set Through the Input of Consumers and Management

A detailed description of this approach will not be given in this chapter, since it is seldom chosen in the "in/out" program. Readers are referred to the material covered in Chapters 3 and 4 (pages 56 and 112 under implementation of the program) to become familiar with the steps and procedures involved in the consumer research needed to collect acceptability responses.

Below is a brief explanation of these steps.

Assessment of Needs and Planning Phase

Involves the identification of needs, resources, and test parameters.

Sample Collection and Initial Screening of a Broad Array of Production Samples

A survey is conducted to collect and screen samples representative of the typical production variability.

Descriptive Analysis

A complete sensory descriptive characterization of the screened samples is obtained. Data are used for ensuing steps of the program, such as sample selection and data analysis.

Sample selection

Descriptive data are analyzed to select the minimum number of samples that span the typical production variability for consumer testing.

Planning and Execution of the Consumer Research Study

The research study is conducted to collect the consumer responses to the product's variability.

Data Analysis/Data Relationships

Consumer and descriptive data obtained above are analyzed to identify relationships between descriptive intensities and acceptability, which are summarized for management to review.

Establishment of Sensory Specifications and Product References

A management meeting is scheduled to review the results. Sensory specifications are established, and product references for training are selected.

Specifications Set Based on Management Criteria Only
This process is explained in detail, as it represents the most popular method for establishing "in/out" specifications. This process is illustrated by using as an example a manufacturer of fruit yogurts that has limited research resources and funds and does not have the support of a R&D sensory group. Therefore, the company has hired a consulting firm to design, implement, and initiate the operation of the program, and to administer the various tests required during the implementation, such as the descriptive analysis.

The company produces five yogurt flavors in only one plant. Historical data show that production variations manifest themselves similarly across all products, except for flavor. Specifically, the character and intensity of off-notes varies across the different fruit flavors. Based on this information, management decides to initiate the program with two flavors and to add the other flavors at a later date. The introduction of new flavors to the program is simple, since the focus is on flavor problems only. The program is initiated for strawberry and vanilla yogurts, since these flavors represent the range of problems found during regular production. All steps described below are followed for the two flavors, although the discussion focuses primarily on strawberry yogurt.

Sample Collection and Screening
A complete survey is scheduled to collect 50 strawberry yogurt production samples from the plant in one season. Another small survey is scheduled for the winter season to assess seasonal fruit variations on the product's variability.

Two sensory consultants, experts in descriptive analysis of foods, and two QC professionals of the yogurt company screen the 50 samples to select products that differ from what is considered "typical" production. Upon the completion of this evaluation, the following is available:

- The documentation of the most variable attributes of strawberry yogurt;
- The documentation of the range of variability expressed in attribute intensities; and
- The selection of samples representative of extreme and intermediate variations.

Figures 5.3 and 5.4 show these results for strawberry and vanilla yogurts. The ranges are indicated on 15-point intensity scales (where 0 = none and 15 = extreme). The distance between the left end marked "0" and first slash mark on the line represents the lower range intensity value. The circles in Figure 5.3 represent the strawberry yogurt samples representative of extreme and intermediate intensities of variable attributes.

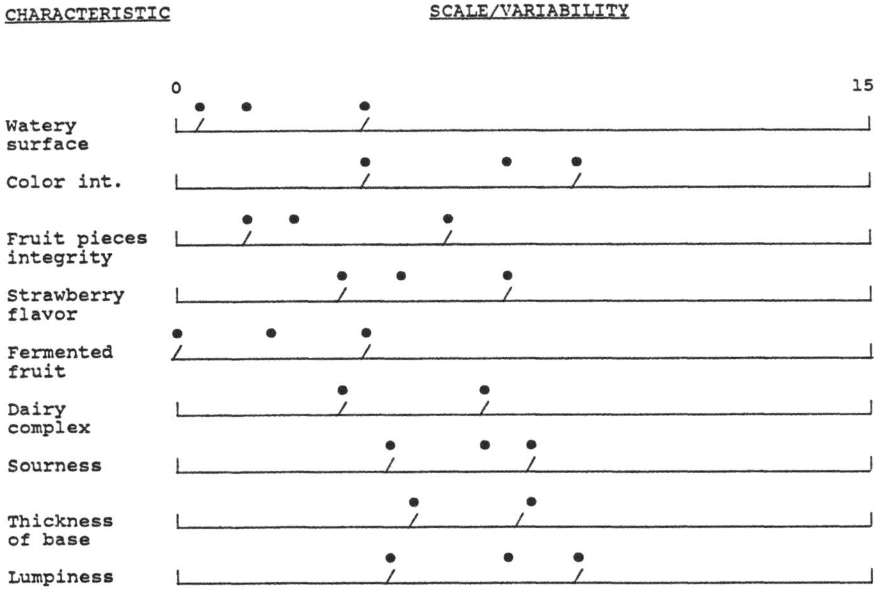

CHARACTERISTIC   SCALE/VARIABILITY

FIGURE 5.3.   Summary of strawberry yogurt production variability and samples selected for the management meeting.

CHARACTERISTIC   SCALE/VARIABILITY

FIGURE 5.4.   Summary of vanilla yogurt production variability.

Establishment of "In/Out" Specifications
A management meeting is scheduled to establish the "in/out" specifications for strawberry yogurt based on the survey results. The range of each variable attribute is illustrated by showing product references that represent the extremes and intermediate intensities (circles, Fig. 5.3).

The following five steps are followed to establish the "in/out" specifications:

1. Selection of the *critical* variable attributes (i.e., those variable attributes considered important by management);
2. Selection of "in/out" limits of each critical attribute (i.e., the minimum or maximum attribute intensity that defines an "in-spec" product);
3. Establishment of the "in/out" specification (i.e., group of all attribute limits and qualitative characteristics that define "in-spec" product);
4. Identification of product references that represent "in-spec" and "out-of-spec" production; and
5. Identification of the action standard that will be used to determine product disposition based on the results of the "in/out" panel.

This process for the strawberry yogurt example is summarized below:

1. Critical attributes. After reviewing the results of the survey (Fig. 5.3), management defines the following to be the critical strawberry yogurt attributes:
   a. Integrity of fruit pieces;
   b. Strawberry flavor intensity (complex);
   c. Fermented fruit; and
   d. Lumpiness.
2. "In/out" limits. The tolerable limits of each critical attribute are set (Fig. 5.5).

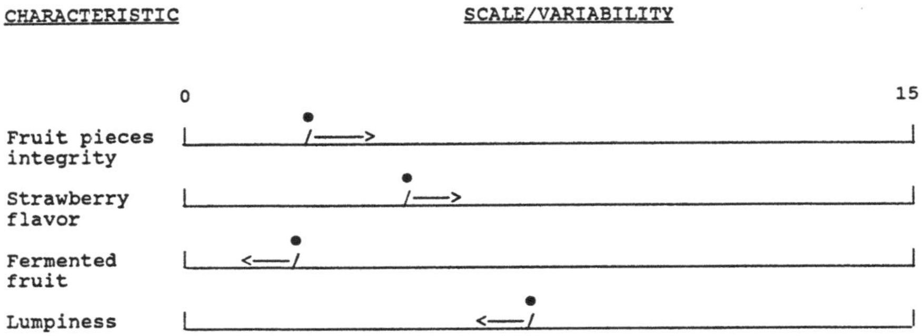

FIGURE 5.5.   Establishment of "in/out" specification and selection of product references.

For example, management considers strawberry flavor intensities lower than 4.5 as unacceptable and sets an "in/out" limit of 4.5 for this attribute. Therefore, products with intensities lower than 4.5 in this attribute are considered "out-of-spec."

3. "In/out" specification. The "in/out" specification represents the space defined by all the "in/out" limits and other "in" or "out" qualitative criteria considered critical for that product. Some qualitative "out" criteria may be: "presence of off notes," "presence of a *different* peppermint character," "off color," "specks," "blotches," and so on. "In-spec" products fall within the space defined by the attribute limits. "Out-of-spec" products fall outside one or more of the attribute limits established in step 2 and have one or more of the qualitative "out" characteristics (e.g., "off-color").

4. Product references. Figure 5.5 shows the product references (circles) selected. These products are to be used for documenting the "in/out specification" and training purposes.

5. Action standard. The output of the "in/out" panel is the percentage of panelists that rate the batch as "in" (denoted as P). Ideally, all appropriately trained respondents should find "out-of-spec" samples to be "out" (i.e., P = 0) and "in-spec" samples to be "in" (i.e., P = 100), yielding the "step-function" plot in Figure 5.6a. However, in practice, the observed values of P vary between 0 and 100, following the sigmoidal-shaped curve presented in Figure 5.6b. If there is no bias in the respondents' evaluations, a sample falling right on the "in/out" boundary would have a 50 percent chance of being evaluated as "in" and a 50 percent chance of being evaluated as "out."

Therefore, management may choose to set the action standard based on "in/out" evaluations as being, "Accept any batch of product for which P ≥ 50% and reject any batch for which P < 50%." Alternatively, in recognition of the variability of the panel responses, management may choose to create a "gray zone" around the cut-off point with an action standard such as, "Accept any batch for which P > 60% and reject any batch for which P < 40% Submit batches for which 40% ≤ P ≤ 60% for additional testing (e.g., descriptive analysis)." The selection of an action standard is a management decision. It is not the sole responsibility of the sensory coordinator. For the strawberry yogurt program, a 50-percent cut-off point for P was selected.

Descriptive Analysis
Several of the product references identified in the management meeting (Fig. 5.5, circles) are evaluated by a descriptive panel managed by the consulting firm. The products evaluated are those samples that represent the "in/out" limits of all critical sensory attributes. For example, to document the "in/out" limits for strawberry

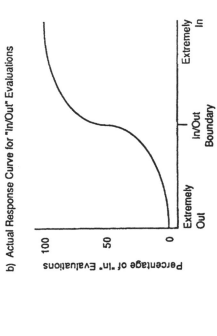

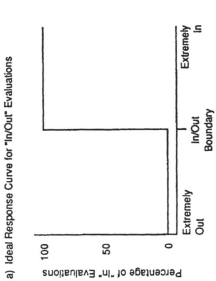

FIGURE 5.6. The ideal (a) and actual (b) response curves for "in/out" evaluations.

flavor intensity, the sample with an intensity of 4.5 is evaluated. A complete appearance, flavor, and texture characterization is obtained for each of the references. Table 5-3 shows the results of two references—the lowest and the highest intensity references for fermented fruit. These results are the sensory baseline documentation of the references needed for the selection of replacement references at later stages of the program. The characterizations of the replacement references

Table 5-3.   Descriptive analysis results of two strawberry yogurt references

	Sample with no fermented	Sample with highest fermented
APPEARANCE		
Watery surface	1.1	0.5
Color chroma	8.2	7.3
Color intensity	6.5	7.0
Fruit pieces/amt	5.2	6.0
Fruit pieces integrity	3.7	2.7
Lumpiness	7.9	6.1
FLAVOR		
Strawberry complex	6.5	4.4
Fresh strawberry	2.5	0.5
Cooked strawberry	4.1	4.0
Fermented fruit	0	2.5
Dairy complex	5.8	4.5
Cultured	4.1	3.4
Milky	2.3	1.5
Butterfat	1.2	0.5
Unripe cultures	0	0
Sweet	8.5	8.0
Sour	6.1	5.0
Astringent	4.3	4.0
TEXTURE		
Lumpiness	7.5	6.0
Firmness	6.7	5.5
Cohesiveness	4.1	3.5
Fruit awareness	4.5	4.0
Mixes with saliva	9.5	9.2
Cohesiveness of the mass	3.3	3.0
Lumpiness of the mass	4.7	3.5
Dairy film	4.2	4.0
Residual fruit pieces	2.6	2.0

are to match the baseline results. References have to be obtained periodically for panel calibration.

Training and Operation of Plant Panels

Panel Selection

Plant employees are recruited to participate in the program. Prescreening and screening phases are completed to select the panel members. Optimally, 25 or more panelists are selected. However, smaller panels are usually formed in small plants. The use of larger panels permits a sensitive analysis of the resulting data, while smaller panels yield highly imprecise "in/out" measurements such that the practical value of the approach comes into question (see "Data Analysis" on pg. 161 in this chapter.).

Sample Selection for Training

The samples used for training are the product references chosen in the management meeting (see "Establishment of 'In/Out' Specifications" on pg. 150 in this chapter). If the training is scheduled immediately after the meeting, perishable products (e.g. foods) collected at initial phases of the program can be used. Less restrictions on scheduling are imposed in nonfood programs.

The same yogurt references collected in the initial survey and chosen by management are used for training purposes, since the training was scheduled immediately after the meeting.

Training Program

The main advantages of the "in/out" training program are its simplicity and short time to complete. The total time required depends on the type of product in the program and varies between 5 and 10 hours.

As in other QC/sensory programs, the two training steps are the basic sensory evaluation phase and the product training phase. In the basic sensory evaluation phase the panelists are familiarized with the applications of sensory evaluation, its importance within the company, and the basic concepts of physiology that play a role in the evaluation of the company's products (e.g., appearance, flavor, texture). The product training phase focuses on teaching the panelists the "in" and "out" product concepts (i.e., the characteristics of "in" and "out" products).

While the comprehensive descriptive (Chapter 3) and quality ratings programs (Chapter 4) require a detailed training in both the qualitative and quantitative aspects of attributes, the "in/out" training program requires no specific attribute training, since no information on product attributes is collected. Only a very general discussion of product attributes is held to allow panelists to focus on those attributes that determine "in-spec" and "out-of-spec" product. The five steps followed in this training program are:

1. A discussion of the nature of "in/out" judgements (i.e., methodology to follow);
2. The identification of the critical attributes that are to determine "in" or "out" production;
3. The review of each "in/out" attribute limit;
4. The integration of all concepts into defining the "in" and "out" product concept; and
5. The evaluation of references and regular production samples (practice period). This process is illustrated for the strawberry yogurt example.

The Nature of "In/Out" Judgements
The product training phase is initiated by familiarizing panelists with the nature of the "in/out" judgements. Visual demonstrations are used to explain this concept, such as space/size and color examples. For example, using cubes of different volumes for this demonstration the "in/out" concept is taught as follows:

• The definition/identification of the characteristic or attribute to focus on (i.e., the volume of cubes);
• The identification and learning of the attribute limits (the specification) that distinguish "in" from "out" product (i.e., the *specific* cube/volume that defines that limit); and
• The demonstration of a series of "in" and "out" products for panelists to practice (i.e., a series of smaller and larger cubes than the "limit cube" (specification) are shown for panelists to judge if they are "in-spec" or "out-of-spec" products).

Critical Product Attributes that Determine "In" or "Out"
Production
Panelists are familiarized with the critical product attributes selected by management. These attributes are to distinguish "in" from "out" production. A definition of each is briefly covered. Table 5-4 shows the information given to the panelists trained in the strawberry yogurt program, where the critical strawberry yogurt attributes are identified.

The Review of Each "In/Out" Attribute Limit
The review of each attribute is completed through a procedure similar to that followed with non-product examples (e.g., cubes), which is:

• A brief definition of the attribute;
• The demonstration of a series of samples that spans the variability of that attribute;

Table 5-4.   List of all strawberry yogurt sensory attributes and critical attributes (*) chosen by management

Appearance	Flavor	Texture
Watery surface	* Strawberry complex	* Lumpiness
Color chroma	Fresh strawberry	Firmness
Color intensity	Cooked strawberry	Cohesiveness
Fruit pieces/amt	* Fermented fruit	Fruit awareness
*Fruit pieces integrity	Dairy complex	Mixes with saliva
Lumpiness	Cultured	Cohesiveness mass
	Milky	Lumpiness mass
	Butterfat	Dairy film
	Unripe cultures	Residual fruit pieces
	Sweet	
	Sour	
	Astringent	

- The identification of the attribute level that defines the limit between "in" and "out" product (i.e., "in/out" specification); and
- The practice of "in/out-of-spec" judgements for that attribute using "in-spec" and "out-of-spec" products. This process is illustrated for one of the strawberry yogurt attributes, fermented fruit, in Table 5-5.

Integration of All Attribute Concepts
Upon completion of the attribute training, all attribute concepts are reviewed and integrated to define the "in-spec" space or concept. The combination of all "in/out-of-spec" limits and the "out" qualitative criteria identified by management (e.g., presence of off-notes, off main flavor character) define this space or concept.

Panelists learn that products falling outside one or more of the "in/out" limits or have one or more of the qualitative "out" characteristics are to be judged as "out-of-spec" To illustrate this concept, a series of products that fall either "in or out" of the specification are shown to panelists in random order. The products are samples reviewed previously during the training on attributes and are identified to the panel as "in" or "out" products. If the panel is to indicate the reasons why a product was judged as "out-of-spec." in the routine evaluations (i.e., as shown in Fig. 5.2), the attribute is also identified in this excercise.

This review reinforces the learned concepts and procedures to be followed in routine evaluation.

Practice Period
One to three practice sessions, as described above, are scheduled after the training is completed. In these practice sessions, samples are only identified with three digit

Table 5-5.   Approach followed in the training of "in/out" judgements per attribute

Attribute: Fermented fruit

1. Definition: The aromatic associated with overripe and soured fruit resulting from fermentation.

2. Samples representative of variability

Code	Intensity Fermented Fruit*
635	0.0
128	2.0
439	4.0

3. "In/Out" limit

Code	Intensity
128	2.0

4. "In-spec" and "out-of-spec" review

Products	Judgement
035	In-spec
128	In-spec
439	Out-of-spec

* Intensities are not presented to panelists. This information is obtained through descriptive analyses and is only handled by the sensory coordinator.

codes. Panelists judge products to be either "in-spec" or "out-of-spec." Results are reviewed with the panel, and problem areas, if any, are discussed. At that time, product references are reviewed, if necessary.

In the last session, it is recommended that the products representative of "in/out" limits of all attributes are reviewed with the panel, even if no problems have been detected. Panelists are instructed that learning and recalling those limits in routine evaluations determines the success of the program. Therefore, references are to be reviewed on a regular basis, or as often as the panel needs the review. Once the practice sessions are completed and problems are resolved, the ongoing operation of the program is started. This means that daily production or incoming ingredients can be judged by the panel.

Panel Maintenance

The "in/out" QC/sensory program also requires a maintenance program that addresses psychological and physiological considerations. Chapter 3 summarizes some of the activities involved in the maintenance program. Among the psychological aspects to be addressed are recognition and appreciation of the panelists participation, planning special group activities, rewards, panel performance feedback, and panel review sessions.

The physiological factors described in Chapter 3 are also applicable in the "in/out" QC/sensory program. These are:

- The scheduling of review sessions. These sessions, scheduled approximately once a month, allow the review of technical and other program parameters. Among the technical aspects to be reviewed are product evaluation procedures and "in/out" limits. Other aspects of the program that need to be addressed are panel scheduling, conditions of the panel room, sample preparation, and so forth.
- The review of "in/out" limits. The review of "in/out" limits in this program is critical. Panelists need to be recalibrated with the products that represent the limits of "in-spec" and "out-of-spec" production to maintain adequate performance.

For some perishable products, a large amount of the products that represent the "in/out" limits can be obtained initially, stored under controlled conditions for a long time, and used as needed. For most products, however, replacement product references need to be produced/identified on a regular basis. To properly administer the acquisition and use of "in/out" references, the sensory coordinator needs to use the descriptive data base gathered at initial stages of the program (pg. 147 or 151), and a descriptive panel that can characterize the sensory properties of the replacement products considered for "in/out" references.

The review of "in/out" references allows panelists to recall the product criteria to judge if products are "in-specification" or "out-of-specification." This practice also decreases within panelist variability.

- The evaluation of blind references. Products known to be "in spec" and others known to be "out of spec" (blind controls) are presented in the set to monitor that panelists mark the samples appropriately. For instance, Table 5-6 summarizes an early panelist monitoring study on the yogurt "in/out" panel. Panelists 5 and 22 are overly critical, marking many "in-spec" samples as "out"; conversely, panelists 21, 25, and 28 are too lax, marking many of the "out-of-spec" product as being "in"; and panelists 11 and 16 exhibit a low level of agreement for both "in-spec" and "out-of-spec" samples. These findings indicate the need to schedule sessions to review the "in/out-of-spec" limits.

The Establishment of a Data Collection, Analysis, and Reporting System

Data Collection
Samples to be evaluated by the plant panel are collected at the same time as the samples that will undergo other QC testing. The frequency of sampling (i.e., the

Table 5-6.   Panelist monitoring results on level of performance with hidden reference samples

Panelist	Percent Correct when Hidden Reference was:	
	In	Out
1	71	82
2	78	76
3	80	77
4	73	75
5	47	91
6	83	79
7	80	81
8	72	76
9	75	79
10	84	81
11	49	46
12	77	73
13	83	80
14	73	76
15	71	74
16	41	47
17	78	72
18	81	83
19	79	81
20	84	85
21	92	48
22	44	91
23	84	71
24	83	76
25	90	48
26	75	77
27	81	72
28	93	49
29	79	81
30	77	78

number of pulls per batch) is determined using the same QC criteria that apply to all other tests. For example, the strawberry yogurt is sampled three times (early, middle, and late) in the packaging step of each production batch.

Unlike other QC/sensory methods, there are situations in the "in/out" method that may require that each panelist evaluate a given sample of product several times. The number of times each panelist evaluates a single sample is determined by the level of precision that the sensory coordinator requires of the data. The issues influencing the sensory coordinator's decision are discussed in detail in the following section on data analysis. For the strawberry yogurt, the sensory coordinator has determined that each of the three production samples (early, middle, and late) must be evaluated four times in order to achieve the necessary level of precision. Four separately coded samples from each of the three production samples are prepared and presented to the panelists, yielding a total of 12 samples to be evaluated. A complete block presentation design is used. The results of the respondents' "in/out" evaluations for batch A3-0028 of strawberry yogurt are presented in Table 5-7. The table is arranged for convenient data analysis that takes into account the multiple evaluations of each production sample.

Table 5.7.   Example of a tally sheet for the "in/out" method for a batch of strawberry yogurt (I = "in," O = "out")

Batch: A3-0028

Sampling Period / Samples / Respondent	Early 928	532	174	825	Middle 692	773	324	418	Late 992	586	273	401
1	I	I	I	I	I	O	O	I	I	I	I	I
2	I	O	I	I	I	O	I	O	I	I	O	I
3	I	O	I	I	O	I	I	I	I	I	I	I
4	I	I	I	I	I	I	I	I	O	I	I	O
5	I	I	O	I	O	I	O	O	·I	I	I	I
6	I	I	I	O	I	I	I	I	I	O	I	I
7	I	I	I	I	I	O	O	O	I	I	O	I
8	I	O	I	O	I	O	I	I	I	O	I	I
9	I	I	I	I	I	I	O	O	I	I	I	O
10	I	I	I	I	I	O	I	O	I	I	O	O
11	I	I	I	I	O	I	I	I	O	I	I	I
12	I	I	O	I	I	O	O	O	I	I	O	I
13	O	I	I	I	I	I	I	I	I	I	I	I
14	I	I	I	I	O	O	I	O	I	I	O	I
15	I	I	I	O	I	I	O	O	I	I	I	I
16	I	O	I	O	O	O	I	O	I	I	I	I
17	I	I	I	I	I	I	I	I	I	I	I	I
18	O	I	I	I	I	I	I	I	I	O	I	O
19	I	I	I	O	I	O	O	O	I	O	I	I
20	I	I	I	I	O	I	I	I	I	I	I	I
21	I	I	I	I	I	O	O	O	I	I	I	I
22	I	I	I	O	I	I	I	I	O	I	O	I
23	O	I	I	I	I	I	O	O	I	O	O	I
24	I	O	I	I	O	I	I	I	I	I	I	O
25	I	I	I	O	I	O	I	O	O	I	I	I
Number "in"	22	20	23	19	18	14	16	12	21	20	18	20
Sample P	88	80	92	76	72	56	64	48	84	80	72	80
Period P	84				60				79			
Batch P	Average = 74.3											
	Range = 24.0											

Data Analysis

The statistical treatment of the "in/out" responses is straight forward. The proportion of respondents who rate the sample as "in" (P) is used to estimate the proportion of all possible respondents who would find the product to be within its standard frame of identity, where:

$$P = 100 * \frac{\text{Number of "In" Ratings}}{\text{Total Number of Respondents}}$$

P is treated statistically in the same way as any continuous QC variable (e.g., ingredient concentrations, attribute intensities).

One of the first things to consider when dealing with a continuous QC variable is the relative precision of the measurement. Simple proportions contain little quantitative information, so, in order to achieve an acceptable level of precision (i.e., an acceptably small standard deviation), relatively large numbers of respondents are required, as compared to other QC/sensory methods. The sample standard deviation of P, based on 25 respondents, is:

$$S = \sqrt{\frac{P(100-P)}{25}}$$

which for P = 50% yields S=10%. This is an extremely large measurement error for a QC method. The precision could be increased by increasing the size of the panel. However, doing so has limited practical value. The number of respondents would have to be increased to 100 to double the precision (i.e., S=5%) versus a 25-person panel. Another way to obtain the same increase in precision is to have each panelist evaluate four samples of product from the same batch (or if within-batch sampling is being done, to evaluate four samples of product from the same sampling period). P would be computed for each sample, and the average of the four P values would be reported as the panel result. The sample standard deviation of the average of four evaluations is half that of a single evaluation (see Appendix 1), since:

$$S_4 = S/\sqrt{4} = S/2$$

In practice, a combination of panel sizing and multiple evaluations needs to be used to achieve an acceptable level of precision.

In the strawberry yogurt example, three production samples are pulled from each batch of product, and four panel evaluations are performed on each sample. The first step of the data analysis is to compute P for each of the 12 samples (called

"Sample P" in Table 5-7). The four P values for each production sample (i.e., early, middle, and late in the batch) are then averaged to yield a "Period P" corresponding to the time period in which the production sample was obtained. The "Period P" in Table 5-7 are the raw QC/sensory data for each batch of strawberry yogurt—that is, the overall batch average P and the range of P values for the batch are computed from the three " Period P " values (see Table 5-7).

The P values should be recorded in a format consistent with that used for other QC tests.

Specifically, if multiple samples are collected within each batch, the batch average and range (i.e., the average and range of the Period P values) are computed and recorded (see Table 5-8). If only a single sample per batch is collected, the single P value (an average, if multiple evaluations are performed) is the batch rating; no measure of within-batch variability is available.

Report System

The sensory coordinator should highlight any samples that fall "out-of-specification," using the company's standard procedures for such occurrences. For example, the management team for the yogurt manufacturer has chosen a 50-percent cut-off point for P . Batches of product with an average P less than 50-percent are rejected. The current batch of strawberry yogurt (A3-0028) has a batch average P = 74.3% and, therefore, is acceptable from the QC/sensory perspective. However, several recently produced batches had unacceptably low P values. The sensory coordinator has highlighted those batches that fail to meet the action standard and submitted the results for management review and action (see Table 5-8). The key factor to bear in mind is that the P should be treated in exactly the same manner as any other analytical measurements being collected for QC. All standard company formats, reporting procedures, and action standards apply.

If an SPC program is in effect, control charts for the P response should be

Table 5–8. Results on several production batches of strawberry yogurt using the "in/out" method

Batch	Percent "In" Average	Range
A3–0016	65.6	20
A3–0018	59.8	19
A3–0020	58.4	18
A3–0024	42.6*	23
A3–0026	44.8*	17
A3–0028	74.3	24

*Out of Specification (Percent "In" 50%)

maintained and forwarded to appropriate personnel. If multiple samples per batch are collected, then $\overline{X}$-charts and R-charts are commonly used to track production (see Appendix 5). If only a single sample is collected per batch, I-charts are used in place of $\overline{X}$-charts. No analogue for R-charts is available when only one sample per batch is collected.

The $\overline{X}$-chart and R-chart for the strawberry yogurt example are presented in Figure 5.7. Note that prior to production of the "out-of-spec" batches A3-0024 and A3-0026, the six previous batches displayed a consistent decreasing trend in P . Based on Nelson's (1984) criteria (see Appendix 5), this observation is sufficient to conclude that the yogurt process was "out of control." The need for corrective action was indicated before "out-of-spec" product was produced. Acting on this observation may have prevented the loss of the two batches.

Characteristics of the Ongoing Program

The activities of the program in operation are grouped into daily/routine and long-term activities listed here (see Chapter 3 for a detailed discussion).

Daily Routine Program Activities

The daily routine program activities include:

1. Scheduling of panel sessions;
2. Scheduling of panelists' participation;
3. Sample and reference preparation;
4. Administration of test;
5. Data analysis, report, and interpretation with other QC data;
6. Scheduling of special sessions; and
7. Panel maintenance.

Long-Term Program Activities

The long-term program activities include:

1. Program growth:
   a. Introduction of new products;
   b. Introduction of other points of evaluation in the process;
   c. Introduction of new critical attributes to be included in the "in/out" guide-lines;
   d. Design and execution of research projects; and
   e. Improvement of sensory methods (modification of the "in/out" method or introduction of new sensory methods).

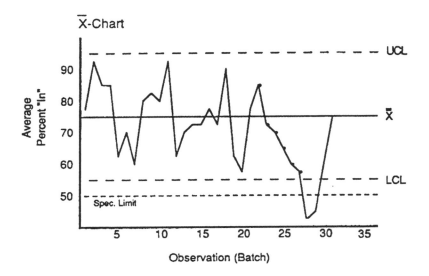

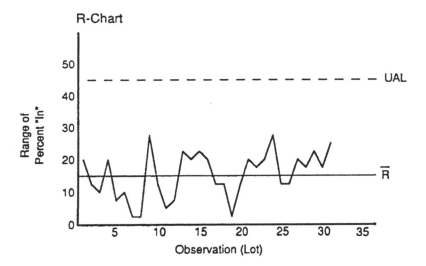

FIGURE 5.7.   The $\overline{X}$-chart and R-chart for batches of strawberry yogurt evaluated using the "in/out" method. Note the six consecutive decreasing batches that indicate that the process was "out of control."

2. Sensory coordinator's development.
3. Panelists' development.

Chapter 3 gives a description of all these activities, procedures, requirements, benefits, and special considerations.

OTHER VERSIONS

"In/Out" Judgement with Descriptive Data

The concepts of a descriptive method (attribute evaluation) can be incorporated into the "in/out" program. The rating of key attributes is incorporated to provide information on specific attribute variations and on the reasons why a given product was considered "out-of-spec" (see Fig. 5.8).

The advantage of this modified program is the collection of additional product information, which is helpful in making decisions on product disposition. The disadvantage is that the program becomes more complex, requires additional training and resources, may increase the time to complete evaluations, and may decrease the number of samples evaluated in a session. Specifically, the panel needs to be trained in the detection and description of attributes, and attribute references need to be collected and maintained.

The incorporation of descriptive concepts into the "in/out" method is not recommended for those cases in which a simple and quick evaluation is needed (e.g., simple raw ingredients). Rather, this modified program may be valuable in the evaluation of more complex products, for which the additional descriptive information is helpful in defining the way production varies. Some examples include scented products (e.g., laundry detergents), lotions and other personal care products, and more complex raw ingredients or food products (e.g., chocolate).

This modified "in/out" program tends to become a descriptive method in the long term, especially as the panel becomes well trained and skilled. As panelists improve their skills in focusing on and rating individual attributes, they may encounter difficulties in integrating all the attribute information on only *one* integrated judgement ("in" or "out of spec").

"In/Out" Judgement Followed by a Descriptive Evaluation

One of the most effective ways to utilize sensory resources within a quality control function is to establish two programs, such as an "in/out" and a descriptive method, concurrently. The benefits of this approach are not perceived at initial stages of the program, since two methods are implemented simultaneously (e.g., two panels are

```
                                        Name_____

                                        Product code_____


   ■  INSTRUCTIONS

   .  Please look at and smell this product and rate the intensity
      of the attribute listed below using the scale

      0  - - - - - > 10
      none             extreme


        Attribute          Intensity

        Chroma             _____

        Floral             _____

        Citrus             _____

        Animal/base        _____

        Plastic            _____


   ■  "IN/OUT" ASSESSMENT

   .  Indicate if the product is "in" or "out" of specification based
      on your assessment of the individual characteristics.
      Mark one.

        "IN"     [ ]

        "OUT"    [ ]
```

FIGURE 5.8.   Evaluation form used for "in/out" judgement with attribute ratings (scented laundry detergent).

trained). However, this approach is advantageous and efficient once the program is in operation.

In this approach, the "in/out" panel operates routinely as described in this chapter Products are judged to be either "in-spec" or "out-of-spec." Products found "out-of-spec" could be put on hold until the second evaluation is completed by the descriptive panel. This evaluation consists of a brief descriptive characterization of the questionable products. For scented laundry detergent, the attributes evaluated could be chroma, floral, citrus, animal/base, plastic, or other attributes.

The advantages of this approach are:

• The two panels are used in the most efficient manner. The least expensive sensory tool ("in/out" panel) is used routinely, while the most expensive sensory tool is used only in special occasions, when required.

- Descriptive information is obtained on rejected products, which helps explain the reasons why that product is "out-of-specification" and provides guidance to correct problems.

6

Difference-from-Control Method (Degree of Difference)

ABSTRACT

The difference-from-control method (also called the degree of difference method) utilizes a standard overall difference method of rating to determine how much any production sample varies from a control product. This method requires a consistent easy-to-hold or easy-to-reproduce control. A panel of subjects is trained briefly to recognize and rate samples that represent an array of product differences from the control. In some cases, attributes can be rated in addition to the overall difference from control. These attribute ratings can be in the form of intensity ratings or difference from control for the attribute.

The main advantage of this method is the simplicity of a single overall rating that allows for rapid decisions in regard to disposition of daily production. The addition of attribute ratings provides additional information as to sources of the differences.

This program's disadvantage is that the overall difference ratings for a production sample does not provide enough information about the source of the differences to allow for the correction of raw materials or process. Even the inclusion of a subset of attributes may be insufficient to track all variability sources.

PROGRAM IN OPERATION

When a difference-from-control method (also called degree of difference) is implemented as a quality control tool, the intent is to track production at one or more plants in terms of the *amount* that any finished product or raw material differs from a control or standard for that product or material. This overall difference test method is employed to measure the degree of difference and accept product that falls below some critical difference value and is closer to the control, and to reject product that falls above the critical difference value.

Panelists are trained to recognize the control, to recognize an array of products, which differ along a continuum from the control, and to rate those differences using a rating scale. Figure 6.1 shows different forms of difference from control rating scales. Table 6-1 shows difference-from-control ratings for four samples. Each score is based on the mean value for 25 panelists. One sample, a blind control, has the lowest rating, which means it was perceived as closest to the control. This is a check of internal validity of the panel and serves as a measure of the effect of asking the difference question (placebo effect).

QC management uses these data to decide on product disposition, based on previously set criteria for the product. For example, if a difference larger than 5.5 on a 10-point scale is considered out-of-spec, the product 819 is held for rework, while products 476 and 623 are released for distribution (Table 6-1). At no time are the panelists aware of the accept/reject criteria for any product.

```
Mark one line to indicate degree of difference from the control.

_____    0-no difference, same as control

_____    2-very slight difference

_____    4-slight difference

_____    6-moderate difference

_____    8-large difference

_____    10-extreme difference


Mark one box to indicate the difference from the control.

       [ ]   [ ]   [ ]   [ ]   [ ]   [ ]   [ ]   [ ]   [ ]   [ ]
no                                                          extreme
difference                                                  difference


Mark the line with a vertical line to indicate degree of difference.

    |_____|
no                                                          extreme
difference                                                  difference


Using the scale below, indicate with any whole number the degree of
difference.  _____

0    = no difference
20   = very slight difference
40   = slight difference
60   = moderate difference
80   = large difference
100  = extreme difference
```

FIGURE 6.1.   Different types of scales for difference-from-control ratings.

Table 6-1.   Example of difference-from-control test results

Sample Code	Sample Identification	Difference from Control Rating*
476	FJ91	4.6
819	GV97	6.7
531	Blind Control	2.2
623	MC17	5.3

* based on mean values for 25 panelists using a 10-point category scale (0–10)

In this chapter, two cases are described, for which the difference-from-control method is recommended. In both cases, the variation from control tends to be along a single underlying dimension. Case 1 involves a product (air freshener) with a control that is consistent across each unit and can be held consistent over a long period of time. Regular production tends to vary from the control through a loss of fragrance intensity accompanied by an increase in the odor of the carrier or "base."

Case 2 involves a heterogeneous product (bran flakes with raisins), for which the control is not uniform from box to box in amount and texture of raisins and uniformity of the flake size.

In order to account for this box-to-box variability for both the control and for any set of production samples, it is necessary to use 1) the equivalent of a blind control (a sample from the same batch as the control), 2) a secondary control (a sample from a different batch of control product), and 3) the production sample(s) (Aust et al. 1985).

In some cases, the QC/sensory program management decides to include some additional ratings on specific attributes in order to have more information on each product tested.

IMPLEMENTATION OF THE PROGRAM

The difference-from-control method is selected on the basis of the following objectives and product information:

1. Overall objective—To develop a QC method for the sensory evaluation of finished product (and/or raw materials) that determines the size of the difference of any sample/product from a well-defined control.
2. Information gathered and special considerations:
   a. Production variability generally is a function of a major underlying dimension.
   b. Products representing degrees of difference along a continuum can be obtained and held for training purposes.

c. Resources to implement a difference from control program are available at each production facility.

The main steps to implement a degree of difference program are:

1. Establishing sensory specifications by development of a difference from control scale with representative samples and testing this array of samples among consumers;
2. Training and maintenance of the difference-from-control panel;
3. Establishment of data collection, analysis, and reporting system;
4. Continuing operation of the program.

Establishment of Sensory Specifications

Since the control represents the target or accepted sample for any product category, the specifications for that product category are set by determining how large a difference from the control is "too large," that is, large enough to be rejected.

The cut-off point on the difference from control scale can be set by 1) using responses from both consumers and management to an array of products that differ from the control or 2) using management responses only. Using the input of both management and consumers is the preferred approach.

Specifications Set Using Consumer and Management Input

This approach used to set QC/sensory specifications involves the following steps:

1. Assessment of needs and planning phase;
2. Collection and screening of an array of production samples;
3. Descriptive and difference evaluations to characterize a subset of samples;
4. Selection of the array of samples to be evaluated by consumers;
5. Consumer test planning and execution;
6. Data analysis to determine the relationship between the descriptive and difference data and consumer data; and
7. Setting of final specifications based on the data analysis results and management input.

These steps are similar to the approach followed for the comprehensive descriptive method (Chapter 3), except for steps 3 and 7, which involve consideration of the size of the difference from control rating as the major criterion for setting specifications.

The two cases presented above (the air freshener and bran flakes with raisins) are used as examples to illustrate the steps. The air freshener has a control, which is very consistent from container to container and over time, and variation in production that corresponds to one underlying dimension—loss of fragrance. The bran flakes and raisin cereal has variability from box to box of the control and production samples and a need for a measure of changes across a few attributes.

Assessment of Need and Planning Phase
(see "Specifications Set Through the Input of Consumers
and Management" on pg. 56 in Chapter 3)
Since the difference from control test is based on the perception and rating of the size of the differences between test and control products, the selection, production, storage, and use of the control are critical elements of this method. The management of the corporation or division, including Marketing, R&D, Manufacturing, and QC, must be in agreement about the method used to choose *the control* for a brand (see "Initial Program Steps" on pg. 48 in Chapter 2, for discussion of selection of control). This is particularly true for heterogeneous products, such as bran flakes with raisins, for which the test method involves multiple control products. The selection and certification of control products should not be the sole responsibility of the sensory coordinator.

Sample Collection and Initial Screening of a Broad Array
of Production Samples
Production samples of air freshener from five plants, two lines, and two shifts, are collected over a two-week period (five days production/week). The 100 samples collected are screened to identify those samples that represent an array of differences from the control. From among the 100 samples, 6 are selected that typify very small to very large differences from the designated control along the major underlying dimension (loss of fragrance intensity). Any samples that represent other types of differences, such as qualitative difference in the lemon character, are noted. For this product, these differences in lemon blend character are monitored and eliminated during the early screening of the fragrances as raw materials. The major production variable contributing to differences in fragrance and base intensities is the capacity of the process to apply fragrance evenly for each product unit.

For the bran flakes with raisins, samples are selected that represent differences in the primary and most frequent source of variability—heat processing of the bran flakes. The dough is cooked and then toasted. Both the processing times and temperatures have an effect on the amount of browned or toasted flavor and the color intensity of the flakes. Occasionally, some burnt or scorched flavor notes are apparent in the samples, which vary considerably from the control. Other samples that represent some differences in the firmness of the raisins are also selected.

It is possible to have panels rate the intensity of specific attributes, as in the

Modified Spectrum Method (Meilgaard et al. 1987), or rate the degree of difference for those attributes, as will be discussed in the bran flakes with raisins example.

Descriptive Analysis
Although the focus of this method is an overall assessment of differences that can be rated by a descriptive panel, descriptive characterization provides added information in terms of:

- Identification of those samples that best characterize differences along the underlying difference dimension. These then become references for difference ratings and the descriptive characterization documents each difference reference.
- Defining the individual sensory characteristics that contribute to overall difference.
- Identification of samples that represent specific characteristic differences for cases, such as the bran flakes, in which some attribute differences are measured.

See "Specifications Set Through the Input of Consumers and Management" on pg. 56 in Chapter 3 for greater detail.

Table 6-2 lists descriptive terms for air freshener, the panel results for the control and the intensity ranges for the production sample in the survey. This information documents the control product and production ranges. Similar descriptions of the six difference references serve as baseline documentation of the sensory characteristics of each degree of difference reference.

The descriptive information (Table 6-3) for the bran flakes and raisin product

Table 6-2. Attributes for characterization of air freshener and intensity ranges for production survey

	Intensity Ratings for Control	Range for Survey
Overall Fragrance	8.9	4.2–9.2
Lemon/citral	5.2	2.1–5.5
Floral	2.6	1.5–2.9
Sweet aromatic/vanillin	1.7	0 –2.0
Overall Base	3.5	3.0–5.5
Waxy/paraffin	2.1	2.0–3.5
Solvent	1.8	1.0–2.5

Table 6-3.   **Descriptive analysis of control and intensity ranges for production survey**

	Ratings for Control	Range for Survey
Appearance		
Uniformity of flake size	8.6	6.1–10.5*
Color intensity	8.2	8.0–10.2
Chroma	4.1	2.3–4.8
Amount of raisins	6.5	5.0–7.8*
Size of raisins	8.1	8.0–8.4
Flavor		
Grain complex	7.5	6.5–8.0
Raw bran	0.8	0 –1.3
Cooked bran	1.5	0 –2.5
Toasted bran	5.5	4.5–7.0
Caramelized sweet	3.7	3.0–4.1
Cardboard	0	0 –3.8
Burnt	0	0 –2.8
Dried fruit/raisin	4.9	4.0–5.4
Sweet	6.9	6.0–7.4
Sour	2.1	1.5–2.5
Metallic	1.3	0 –2.2
Texture—flakes		
Roughness	10.1	9.0–10.5
Firmness	3.7	3.0–4.1
Crispness	5.2	4.0–6.0
Moisture absorption	7.3	7.0–8.1
Moistness of the mass	8.9	8.5–9.2
Graininess of mass	9.7	9.1–9.9
Tooth pack	4.4	3.4–4.9
Residual cereal pieces	3.9	3.2–4.3
Texture-Raisin		
Firmness	5.1	4.0–7.1*
Cohesiveness	10.4	10.0–11.2
Moistness	12.3	10.0–12.9

* Attributes with a 3-point range

in Case 2 provides the documentation of the control and references for range of the difference as in Case 1. In addition to this information, which describes the differences in the major underlying dimension (effect of heat processing on the flakes), the descriptive data provides information regarding the references that describe differences in other specific attributes. These other attributes include uniformity of flake size, amount of raisins, and raisin firmness. The asterisks in

Table 6-3 indicate that these are characteristics that vary more than three scale points of the rating scale.

Sample Selection
The analyses used for sample selection are described in "Sample Selection" on page 60 in Chapter 3. Some additional analyses can be done to account for the importance of the difference from control ratings.

Following these steps, 6 products are selected for the air freshener consumer test and 13 products are selected for the cereal consumer test.

Planning and Execution of the Consumer Research
A consumer research study, designed to assess the relationship between the overall difference ratings and overall consumer acceptance, and, in Case 2, attribute difference rating and overall consumer acceptance plus attribute consumer acceptance is conducted among consumers who are product users. On page 71 of Chapter 3, the consumer research test design and execution are discussed.

Data Analysis/Data Relationships
The two sets of data collected in step 3 (descriptive and difference) and step 4 (consumer acceptance) above are analyzed to determine relationships between consumer responses to overall differences from the control and differences from the control on specific attributes.

The relationship between overall acceptability and overall difference from control is of primary concern. The relationship between the two responses can be established using regression analysis (see "Data Analysis/Data Relationships" under "Establishment of Sensory Specifications" in Chapter 3) or, for simple relationships of the type presented in Figure 6.2 for the air freshener example, by simply connecting the observations with straight lines. Then, given an action standard for minimum overall acceptance, a maximum acceptable difference-from-control value can be specified. Complex products (i.e., those that exhibit independent variability in several perceivable dimensions) typically require that, in addition to overall ratings, the relationships between consumer acceptance for particular attributes and the difference-from-control ratings for those attributes need to be established to set product specifications. The same techniques for determining the nature of the relationships and for setting cutoff points, as used for the overall ratings, are also used for the individual attributes. The attributes should be grouped into sets containing those that exhibit no relationship between acceptance and degree of difference, those that exhibit simple, typically negative-linear relationships, and those that exhibit more complex relationships (see Fig. 6.3 for examples of each type of relationship). In the meeting with management,

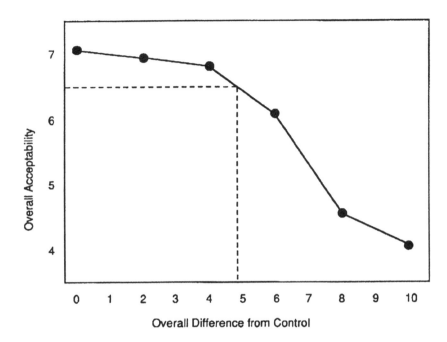

FIGURE 6.2.   Overall difference-from-control vs. overall acceptability (air freshner).

the presentation of the relationships should proceed from the simple to the complex.

Establishing Final Specifications
Once the data analysis and interpretation are completed, management meets with the QC/sensory personnel to establish final sensory specifications. For the difference from control method, the meeting objectives include:

- To review the interpretation of the data in terms of the relationship between overall difference and liking, and, when applicable, between difference for an attribute and liking for that attribute; and
- To set specifications for overall difference ratings, and, where applicable, difference ratings for attributes.

Review of Data Relationships   Products for which difference, descriptive, and consumer data were collected can be presented to management. The products should represent the *range* of *difference*, from a blind control to the most different

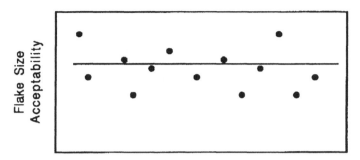

Difference from Control Unif. of Flake Size

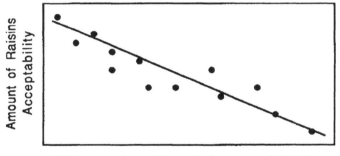

Difference from Control Amount of Raisins

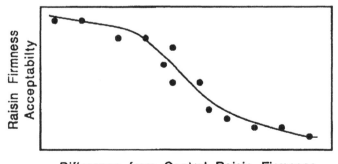

Difference from Control Raisin Firmness

FIGURE 6.3.   Difference-from-control for attributes vs. consumer acceptance for attributes.

sample (10). The relationship of these samples, their difference ratings, and their effect on consumer responses are discussed.

Management may chose to include some attributes that it considers critical or essential to product integrity and that were not high in variability and/or did not have dramatic effect on consumer responses. Alternatively, management may chose to eliminate certain attributes (uniformity of flake size) in Case 2—bran flake and raisin cereal—because they do not have the capacity to control the variable.

Setting Sensory Specifications For the difference-from-control method, management's primary consideration is the relationship between consumer acceptance and overall difference from the control. In both cases (air freshener and bran flake and raisin cereal), a critical value for consumer acceptance needs to be selected, based on past consumer research data.

In the case of the air freshener, Figure 6.2 shows that the critical value of 6.5 in consumer acceptance is considered a critical value. This value corresponds to a difference rating of 5.0. Having reviewed the samples representing 2, 4, 6, 8, and 10 differences, management decides to set a 4.1 as the difference rating that is to be considered too large to ship, the reject point.

Discussions regarding which attributes to include are based on several factors. For uniformity of flake size, the size of difference does not influence liking, and that attribute is dropped. Since raisin counts are done regularly in the QC lab, the "amount of raisin" attribute is eliminated from the sensory evaluation. Specification is set for difference ratings on raisin firmness, since it is a variable attribute and affects consumer responses. In addition, the attribute of flake chroma is to be included, since the descriptive intensity data is highly correlated with overall liking (Fig. 6.4). Difference from control references can be developed from the samples in the consumer research study. In addition to deciding on the critical difference and any additional attributes to be evaluated, management should also confirm the array of samples that represent the various degrees of difference from the control. These samples then can be used as training references.

Final Evaluation Form After the management meeting, the training program is scheduled and the evaluation form to be used in product evaluation at the plants is developed. Figure 6.5 shows the evaluation form to be used to determine degree of difference for the air fresheners. In Figure 6.6, the form used for overall difference and selected attribute difference are shown for bran flake and raisin cereal.

Specifications Set Using Management Input Only
In the absence of consumer responses from the previously mentioned consumer research, it is possible for management to set sensory specifications from the information in the product survey. This survey includes not only an array of

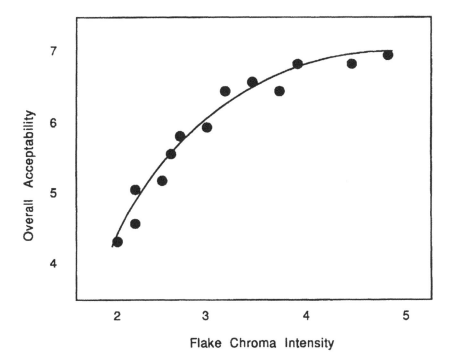

FIGURE 6.4.   Acceptance vs. attribute intensity (flake chroma).

samples that vary to different degrees from the control, but also, in some cases, samples that represent variations in specific attributes. After evaluating this product array and the corresponding difference ratings, management determines the critical point (the degree of difference from the control) above which product is considered out of specification.

Training and Operation of Plant Panel

Training Steps
The operation of a difference-from-control method requires the selection and training of a panel to evaluate production or raw materials in relationship to the sensory properties of a control.

The panel may be selected from among employees or from residents of the local community. Advantages and disadvantages of each approach are discussed in "Training and Operation of Plant Panels" on page 84 in Chapter 3.

Instructions:

. Test samples in the order shown below.

. Test the control first before each coded sample.

. Uncap the container for 1 to 2 seconds and sniff lightly.

. Rate the difference in overall fragrance from the control.

. Use the scale shown below.

Sample Code   Degree of Difference-from-Control

_____   _____

_____   _____

_____   _____

_____   _____

     Degree of Difference Scale

     0 = no difference
     1
     2
     3
     4
     5
     6
     7
     8
     9
     10 = extreme difference

FIGURE 6.5.   Final evaluation form difference-from-control—air freshner.

The selection of candidates for acuity screening is based on a prescreening questionnaire administered to determine interest, availability, and some basic ability with scales (Meilgaard et al. 1987). Since the task of a difference-from-control test is to *detect* and *describe the size* of the difference between a control and a series of production samples, the acuity screening should involve tests for these two tasks.

Candidates are asked to participate in a series of three triangle tests (Meilgaard) in which the control is paired with production samples representing large to small differences from the control. Pairs of samples (control + very different sample, control + moderately different sample, and control + slightly different sample) are

Instructions:

. Test samples in the order shown below.

. Test the control first and before each coded sample.

. Rate the attribute differences for one sample.

. Then rate the overall difference from control for that sample.

. Use the rating scale shown below.

Difference-from-Control

Sample Code	Flake Chroma	Raisin Firmness	Overall Diff.
_____	_____	_____	_____
_____	_____	_____	_____
_____	_____	_____	_____
_____	_____	_____	_____
_____	_____	_____	_____

```
0 = no difference

2 = very slight difference

4 = slight difference

6 = moderate difference

8 = large difference

10 = extreme difference
```

FIGURE 6.6   Final evaluation form difference-from-control—bran flake with raisin cereal.

also presented to the candidates for rating, using a difference-from-control scale (see Fig. 6.2 for scale choices). Candidates who can select the odd sample in two of the three triangle tests and who can rate relative differences using the difference-from-control scale are selected.

One of two types of panels are used with the difference-from-control method: 1) a large (30 to 40 member) panel that is trained to rate difference from the control or 2) a smaller (10 to 15 member) panel that is more highly trained. The larger panel is more appropriate when only the degree of difference from the control is to be evaluated and rated. When additional information about specific attributes (either intensity or difference from control ratings) is collected, the panel requires more training. As training increases the precision of the panelists, a smaller group can be used.

In the product training phase, panelists are trained to recognize and rate a variety of differences from an identified control that is kept as constant as possible throughout both the training and testing phases.

Sample Selection for Training
The samples used for training are the different production samples used in the consumer study (pg. 171) and final management meeting (pg. 176) to represent the difference array. If the product has a short shelf life (less than three months), an additional product survey may be necessary to collect new training samples.

For the air freshener example, once samples are identified as control or reference samples, they are held frozen for several months with no effect to the fragrance level or quality. Therefore, products stored and held from the initial survey, which represent the differences from control references, are used. The control and references for the cereal product can be frozen, as well. In cases where freezing the control and references is not feasible or when the shelf life of the product is shorter than the time between the initial survey and the training, a new product survey is necessary to identify new references for training.

Training Program
The training program for the difference-from-control method involves two stages:

- A basic sensory phase that introduces the panelists to basic sensory evaluation definition, physiology, and applications of sensory techniques within the company in general and with respect to quality measures at the plants.
- The difference from control method, which includes learning to detect and rate the size of the differences from the control product overall, and, in some cases, for specific attributes. This stage also includes training for any special evaluation techniques (sniff, chew, manipulate) that will insure consistency across panelists.

For the air freshener example, the product training phase involves familiarizing the panel (30+) with the product control. One essential aspect of the training process is to provide the panelists with enough exposure to the control, so it is easily recognized. Therefore, the first and primary step is to introduce the control product and provide some sensory cues to the panelists to improve on both recognition and sensory memory for the control. Table 6-4 shows the information given to the panel, along with the sample to aid in the learning process.

After the panelists have had an opportunity to become somewhat familiar with the control and the evaluation procedure, difference from control references can be introduced in pairs along with the control to demonstrate the types and size of differences that panelists may be expected to encounter (see Table 6-5) in normal

Table 6-4.   Training descriptions for control: air freshener

Sample	Control
Procedure for evaluation: Hold container 1 inch from nose; remove cap for 1 to 2 seconds only; sniff contents of container using shallow sniffs. Use this procedure for all samples.	
Difference-from-Control = 0	
Total Fragrance	moderate to strong
Lemon	
Floral	
Vanillin/sweet	
Base Odor	very slight
Waxy	
Solvent	
Difference-from-Control Scale	
0 = no difference	
2	
4	
6	
8	
10 = extreme difference	

product evaluation. The descriptions are not intended to teach the panelists to be descriptive but are intended to provide cues to detecting and remembering what types of differences correspond to certain difference ratings.

After the panelists have evaluated several "known" references with identified difference ratings and descriptive cues, they are given a series of unknowns to compare to the control and rate for overall difference. Included in this series of "test samples" should be a blind control that acts to measure the placebo effect in the difference from control test. The placebo effect is a measure of the degree of difference that the experimenter gets simply from asking the difference question. The real test of difference is the ability of the panel to separate any reference (different) sample from the blind control. (See "Establishment of a Data Collection, Analysis, and Report System" on pg. 189 in this chapter for a discussion of data analysis for degree of difference.)

The training process involves providing enough known references and controls to the panelists so that eventually they are able to feed back the same or almost the same ratings for those references and the control itself, when they are presented as coded samples to be evaluated.

The bran flake and raisin cereal requires more training because evaluation includes difference-from-control ratings for two attributes [flake chroma and raisin firmness], plus the overall difference rating, discussed earlier for the air freshener.

For the cereal, the overall difference ratings are based on differences that are

Table 6-5.   Training descriptions for control and samples

Sample	Difference-from-Control	Sensory Descriptors	Intensity
Control	0	Total Fragrance Lemon Floral Vanillin/sweet	moderate-strong
		Base odor Waxy Solvent	very slight
426	4	Total Fragrance Lemon Floral Vanillin/sweet	moderate
		Base odor Waxy Solvent	slight
527	6	Total Fragrance Lemon Floral Vanillin/sweet	slight-moderate
		Base odor Waxy Solvent	slight-moderate

Difference-from-Control
0 = no difference
2
4
6
8
10 = extreme difference

related to the major underlying variables in the processing of the bran flakes. The overall differences are manifested in increases in toasted flavor and color intensity and occasionally some burnt or scorched flavor notes. Once the panelists are able to provide accurate overall difference ratings for the cereal references, they are introduced to the references and difference ratings for the flake chroma and raisin firmness attributes. Tables 6-6 and 6-7 are examples of the training forms used to introduce each control and two difference references. These include 1) the description intended to help the identification and learning and 2) the procedure used for evaluation.

Table 6-6.   Difference-from-Control—flake chroma

Procedures:
- Use a 75-watt incandescent bulb directly over the samples at 3 to 4 foot distance.
- Sit or stand so the eyes are 2 feet from the sample horizontally and 1.5 to 2 feet above the table surface.
- Set the control valve and test bowl side by side under the light source.
- Shut off the special bulb between sample chroma ratings.

Sample	Difference-from-Control	Description
Control	0	Golden brown No gray color
661	3.5	Less golden brown Some gray tones
219	5.7	Brown Gray tones

Table 6-7.   Difference-from-Control—firmness of raisins

Procedure
- Compress one whole raisin between molars partially.
- Repeat with another whole raisin and compress fully.
- Chew down raisin until disintegrated.

Sample	Difference-from-Control	Description
Control	0	Soft Moist Chewy Slightly springy
113	4.2	Firm Chewy Not springy Slightly moist
708	6.7	Tough Dry

As with overall difference, the reference and control are then repeated as coded samples. Panelists are expected to rate each sample with the difference rating it was assigned initially.

Note: At no time are panelists ever told the critical values of difference for the sensory specifications.

Training for each aspect (overall difference or each attribute difference) can require four hours of training (in addition to the basic sensory training of four hours). Another three to four weeks of practice sessions (one-half hour, three days a week), for a total of about five to six hours of practice, are necessary to insure the ability of each panelist to provide the appropriate rating for the coded (blind) control and references samples. Regular feedback to the panelists, including tabular and graphical summaries of their performances against the known values of the references, should be provided throughout the training/practice period.

For the air freshener example, the average difference-from-control ratings of the reference samples are displayed for ten of the trainees in Table 6-8. The sensory coordinator should examine each panelist's performance versus the known values of the reference samples in terms of both their average ratings and their reproducibility (i.e., their standard deviations). Panelist performance can also be monitored with graphical data summaries. For instance, Fig. 6.7 shows several possible outcomes of a set of panelist ratings of two training samples of the bran flake with raisin cereal. The significance of these patterns is assessed statistically by studying the judge-by-sample interaction effect in an analysis of variance performed on the trainees' data (see Appendix 3). In plot a, all of the panelists agree on the general trend, but differ slightly in their average difference-from-control ratings of the two samples. In plot b, all of the panelists exhibit the same directional trend, but panelists 3, 6, and 7 show a much larger difference between the two samples than the rest of the panel. In plot c, panelists 2 and 7 differ substantially in the pattern of their ratings compared to the other panelists, while panelist 9 shows no difference between the samples. Each of the plots indicates the need for continued training. Plot c is the most severe case, followed by plot b, with plot a representing a minor problem that additional calibration to reference samples should eliminate.

Panel Maintenance
In order to assure that the difference-from-control technique continues to properly function over time, a maintenance program is necessary. Attention is given to both the physiological and psychologic factors discussed in Chapter 3. Psychological considerations, such as panel recognition, rewards performance feedback, and routine panel review sessions, address attitudinal aspects of the panel and panelists.

The physiological maintenance addresses performance aspects of the program and focuses on providing the necessary calibration sessions for panelists to maintain their abilities to detect and rate differences in compliance with the training references. Panelists are told when a blind control or reference appears during regular sample evaluation.

Occasionally, a series of samples from recent production, which have been carefully selected to represent two or three size differences from the control, plus

Table 6-8. Panelist performance summary table—average and standard deviation in degree of difference ratings on unidentified reference samples

Actual DOD	Panelist								Panel
	$\overline{X}\pm S$ (1)	$\overline{X}\pm S$ (2)	$\overline{X}\pm S$ (3)	$\overline{X}\pm S$ (4)	$\overline{X}\pm S$ (5)	$\overline{X}\pm S$ (6)	$\overline{X}\pm S$ (7)	$\overline{X}\pm S$ (8)	$\overline{X}\pm S$
2	2.1±1.3	2.2±1.4	1.9±1.5	2.9±1.7	2.1±1.7	2.6±1.9	2.3±2.0	1.8±1.2	2.2±1.6
4	3.7±1.4	3.8±1.9	4.2±1.5	6.1±2.1	4.1±1.8	3.6±1.6	3.8±1.7	3.7±1.7	4.1±1.9
6	6.0±1.2	6.4±2.0	5.9±1.6	4.9±1.9	6.1±1.4	6.2±1.8	5.9±1.5	6.1±1.5	5.9±1.7
8	7.5±1.8	8.3±1.6	8.2±1.6	7.9±2.0	8.4±1.9	9.2±2.1	7.9±1.3	8.5±1.6	8.2±1.8
10	8.8±1.7	9.1±1.5	9.4±1.3	8.9±1.9	8.9±1.7	9.3±2.0	9.4±1.8	9.6±2.0	9.2±1.8

Entries are : Mean ± Standard Deviation

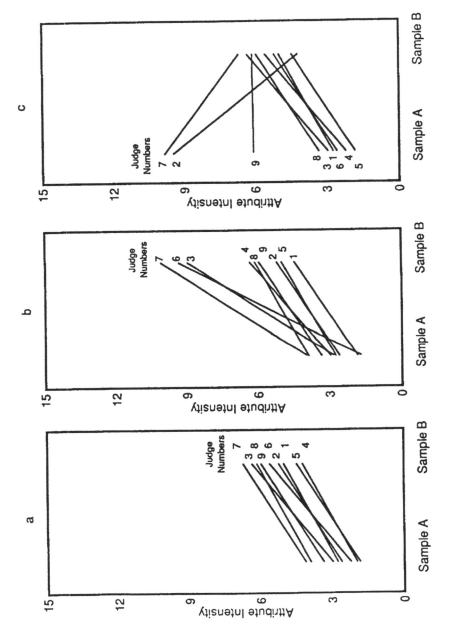

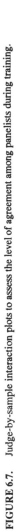

FIGURE 6.7.   Judge-by-sample interaction plots to assess the level of agreement among panelists during training.

a blind control, are submitted for repeated evaluations. Data from the three or four samples [blind control + two or three different rated references] are evaluated for panelist and panel performance. Using the same tabular and graphical summaries that are used during the training/practice period, the panel leader monitors the panelists. These summaries are presented to panelists to inform them of their performance.

In addition, I-charts (see Appendix 5) of each panelist's difference-from-control ratings (both overall and, when appropriate, for individual attributes) on hidden reference samples should be maintained. Biased ratings and/or excessive variability are readily apparent in such charts.

These data summaries serve as maintenance tools for panelists' attitudes as well as daily performance. Management and the QC/sensory analyst and coordinator also have a measure of the panel's and individual panelist's ability.

Establishment of a Data Collection, Analysis, and Report System

Data Collection

Samples to be evaluated by the difference-from-control panel are collected at the same time as the samples that will undergo other QC testing. The frequency of sampling is determined by using the same QC criteria that apply to all other tests. For the air freshener, three production samples are pulled from each batch. For the bran flakes with raisins cereal, only a single sample of product is collected for evaluation from each batch.

The number of samples submitted to the panel in one session is affected by two factors:

1. The number of QC samples that are pulled from each batch of production; and
2. The existence of a substantial amount of perceivable variability among control products.

At each session, the panelists are presented with an identified control, along with a set of unidentified "test" samples (i.e., samples labeled with random three-digit codes). The "test" samples consist of any QC production samples awaiting evaluation and at least one unidentified sample of control product. In the air freshener example, the difference-from-control panel is presented with an identified control and four test samples—that is, the three QC samples that were pulled from the current production batch and one hidden control. There is no substantial variability with control air freshener product. In the bran flake with raisins cereal example, each panelist is presented with an identified control and three test samples—that is, the one QC sample from the current batch and two hidden controls. The hidden controls are obtained from two different batches of control product so as to capture any within-control variability that exists (see Aust et al. 1985). Typically, one of the hidden controls is obtained from the same batch as the identified control; the other is obtained from a different batch of control.

If the number of samples per session is small, a complete block presentation design is used. If, however, the number of samples per session exceeds that which can be evaluated before sensory fatigue sets in, then a modified balanced incomplete block (BIB) presentation design is used. In the modified BIB, all hidden controls are presented in every set of samples (i.e., in each "block"), and the QC samples are presented according to a BIB design (see Gacula 1978). Sensory coordinators using this approach need to consider the number of blocks in the BIB portion of the design because of its impact on the number of panelists required. Further, the sensory coordinators need to be aware that special data analysis techniques (those available, for example, in SAS® PROC GLM) need to be used to summarize the data.

Complete block presentations are used for both the air freshener and bran flakes with raisins cereal examples. In both cases, each panelist evaluates the test samples and rates each with a difference-from-control score corresponding to the perceived difference from the identified control. Only overall difference from control is evaluated on the air freshener samples. For the bran flake with raisin cereal, the ratings include both overall difference-from-control and difference-from-control on specific, critical attributes. A large pool of respondents is used to evaluate the air freshener samples with the current panel consisting of 25 panelists. Their difference-from-control ratings for the hidden control and the three QC production samples are presented in Table 6-9. A smaller, more highly trained group of respondents make up the QC/sensory panel for the bran flake with raisin cereal. For the current batch of cereal, eight panelists evaluate the two hidden controls and the one QC production sample for overall difference from the identified control, as well as difference-from-control on the attributes flake chroma and raisin firmness. The panelists' ratings for the three test samples are presented in Table 6-10.

Data Analysis

The first step in the analysis of data from a difference-from-control panel is to compute the average difference-from-control values for each of the test samples (i.e., both QC production samples and any hidden controls) for both overall difference and any attributes difference scales used. The four sample averages for overall difference of the air freshener samples are presented at the bottom of the panel data in Table 6-9. The sample averages for both overall and attribute differences for the bran flakes with raisin cereal samples are presented similarly in Table 6-10. For both of the examples, simple arithmetic means are computed

Table 6-9.   Difference-from-Control panel data on three production samples of air freshener

Panelist	Hidden Control	Production Sample 1	Production Sample 2	Production Sample 3
1	2	5	4	4
2	3	3	5	3
3	1	2	2	4
4	1	5	3	2
5	0	4	2	1
6	2	4	4	3
7	0	3	3	1
8	1	2	3	2
9	2	3	5	4
10	2	2	1	2
11	3	3	4	5
12	1	3	3	3
13	2	4	3	3
14	1	2	2	4
15	1	2	4	2
16	2	4	3	4
17	0	2	2	2
18	1	5	4	3
19	2	3	2	2
20	1	3	3	3
21	2	3	4	3
22	1	2	4	4
23	1	3	2	2
24	2	3	3	2
25	2	2	3	4
Sample Mean	1.4	3.1	3.1	2.9
Δ		1.7	1.7	1.5
Batch Δ				
Mean : 1.6				
Range : 0.2				

because a complete block presentation scheme is used, then adjusted (least-square) means are used to summarize the panel data.

The next step of the analysis is to eliminate the placebo effect (i.e., the effect of asking the difference question) from the difference-from-control ratings of each of the QC production samples. In situations where within-control variability is not an issue, such as in the case of the air fresheners, this is done by simply subtracting

Table 6-10. Difference-from-Control panel data for a production sample of bran flake with raisin cereal

	Overall			Flake Chroma			Raisin Firmness		
	Hidden Control		Production Sample	Hidden Control		Production Sample	Hidden Control		Production Sample
Panelist	1	2		1	2		1	2	
1	1	2	2	0	3	3	0	1	3
2	0	1	5	0	1	5	0	0	2
3	0	1	3	0	1	5	1	2	2
4	0	0	2	0	1	4	0	1	4
5	0	1	2	1	2	5	0	3	3
6	0	2	4	0	2	6	0	2	4
7	1	2	4	0	2	5	0	1	3
8	0	3	3	0	1	5	0	2	3
Sample Mean	0.3	1.5	3.1	0.1	1.6	4.8	0.1	1.5	3.0
Control Mean	0.9			0.9			0.8		
Batch Δ	2.2			3.9			2.2		

the average of the hidden control sample from the averages of each of the QC production samples, as shown toward the bottom of Table 6-9. In situations where within-control variability needs to be considered, such as in the case of the bran flake with raisin cereal, this is done by first computing the average of the two hidden control samples and then subtracting this average from the averages of each of the QC production samples evaluated in the session—that is:

$$\Delta = \overline{X}_{QC} - (\overline{X}_{C1} + \overline{X}_{C2})/2$$

as shown toward the bottom on Table 6-10 for each of the three scales used to evaluate the cereal samples. In situations where multiple hidden controls are used, the adjustment shown not only eliminates the placebo effect, it also adjusts for the heterogeneity of the control product by, in effect, computing the difference between the QC production sample and an "average" control.

The Δ values are the raw QC data for each production sample evaluated, using the difference-from-control method. If multiple samples per batch are collected, as is done in the air freshener example, the batch average and range (i.e., the average and range of the panel Δ values) are computed and recorded (see Table 6-11). If only a single sample per batch is collected, as in the case of the bran flakes with raisins cereal, then the panel Δ value is the rating for the batch; no measure of within-batch variability is available (see Table 6-12).

Table 6-11.   Summary of Difference-from-Control ratings for the air freshener product

Batch	Overall Difference-from-Control (Δ)	
	Mean	Range
011-E4	1.8	0.8
012-E4	2.2	0.3
013-E4	3.2	0.6
014-E5	1.6	0.2

Table 6-12.   Summary of Difference-from-Control ratings for bran flake with raisin cereal

Batch	Difference-from-Control (Δ)		
	Overall	Flake Chroma	Raisin Firmness
52-0028	3.3	1.3	3.1
52-0029	3.1	2.0	3.5
52-0030	2.5	2.2	2.0
52-0031	2.2	3.9	2.2

In the classical sensory applications of the difference-from-control method, a test of hypothesis (see Appendix 2) is used to determine if production samples are "significantly" different from the control. Aust et al. (1985) modified the test to handle the situation where multiple hidden controls are evaluated. In the QC application of the difference-from-control method, it is recognized that production batches can be considered acceptable even when some level of perceivable difference from control exists, so the classical statistical tests do not apply. Rather, the Δ values are compared to the maximum acceptable difference-from-control specifications (e.g., overall Δ_{max}=4.0 for air fresheners). The sensory coordinator highlights any out-of-spec production in his or her report to management.

Both the air freshener and cereal manufacturers have SPC programs in place so that the sensory coordinators at each company also maintain the control charts of the difference-from-control ratings that are measured on the finished products. For the air freshener product, an $\overline{X}$-chart and an R-chart are maintained for the overall Δ values because multiple QC samples are evaluated from each batch (see Figure 6.8). No indications of an out-of-control process occur in the charts (see Appendix 5). For the bran flakes with raisins cereal, I-charts are maintained for overall Δ, as well as for the Δ values for flake chroma and raisin firmness. I-charts are used because only a single QC sample of cereal is pulled from each batch (see Figure 6.9). Once again, there is no indication of an out-of-control condition in any of the three I-charts. (Applications of control charts, including examples of how to compute upper and lower control limits, are presented in "Data Analysis" pg. 94 in Chapter 3.)

Report System

The sensory coordinator should highlight any samples that fall out of specification, using the company's standard procedures for such occurrences. Additionally, if an SPC program is in effect, the control charts for each of the Δ responses should be updated each time a new batch of product is evaluated, as in Figures 6.8 and 6.9. The updated charts should then be forwarded to appropriate personnel. The key factor to bear in mind is that the QC/sensory data should be treated in exactly the same manner as any other analytical measurements being collected for QC. All standard company formats, reporting procedures, and action standards apply. Once in operation, a successful QC/sensory function should be a fully integrated component of the overall QC program.

Characteristics of the Ongoing Program

As discussed in greater detail in "Characteristics of the Ongoing Program" on page 98 Chapter 3, the ongoing QC/sensory program has daily or routine activities and long-term activities, which are listed here.

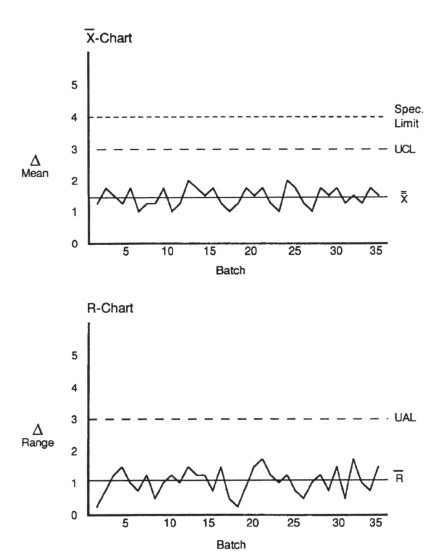

FIGURE 6.8. $\overline{X}$-chart and R-chart of the overall difference-from-control ratings for the air freshner process.

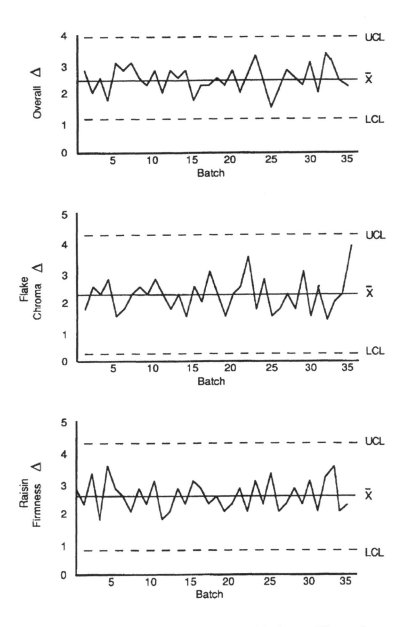

FIGURE 6.9.   I-charts of the overall, flake chroma, and raisin firmness difference-from-control ratings for the bran flake with raisin cereal process.

Routine Program Activities
Regular daily or weekly program activities are:

1. Scheduling of panel sessions;
2. Scheduling of individual panelists;
3. Sample and reference preparation;
4. Administration of test;
5. Data analysis, report, and interpretation, with other QC data;
6. Scheduling of special sessions; and
7. Panel maintenance.

Long-Term Activities
Long-term development activities are:

1. Program growth:
   a. Addition of new products;
   b. Inclusion of other evaluation points in the process;
   c. Design and execution of research projects; and
   d. Improvement of sensory method.
2. Development of sensory coordinator.
3. Development of panelists.

OTHER VERSIONS

As with other sensory methods, some variations of difference-from-control method are possible by combining the difference-from-control rating with ratings for attributes.

Modified Spectrum Method

For Case 2, involving bran flakes and raisin cereal, a few attributes were evaluated in addition to the overall differences from control. In the case described, the attributes also were evaluated in terms of difference from the control. Another alternative described by Meilgaard, Civille, and Carr combines the overall difference-from-control rating with intensity ratings for specific key attributes.

These attributes are selected by the same system used in the Comprehensive Descriptive Method (Chapter 3), and specifications can also be determined, as is described in that chapter.

The benefit of such a method is to provide both 1) data on the variability of specific key attributes that are helpful in determining sources of variability and 2) an overall single rating that is used to determine if the sample is in-spec or out-of-spec "overall" for routine product exposition.

With Full Description

In some products, such as coffee, beer, tea, chocolate, and wine, the flavor attributes are so complex that they require sophisticated descriptive training for all attributes to be evaluated. The resulting method provides the overall difference-from-control rating, upon which decisions about product being in-spec or out-of-spec can be made. The additional full Flavor Spectrum provides detailed flavor documentation in terms of attributes and their intensity. From the full Flavor Spectrum, QC/sensory and R&D can relate the resulting flavor changes to corresponding changes in raw materials and/or processing. An example of evaluation of two coffee production samples and the corresponding control is given in Table 6-13.

Table 6-13.   **Difference-from-Control and full Flavor Spectrum descriptive analysis for coffee**

Attributes/ Samples	Control	FJ91	GV97
Difference from control	0.0	4.6	6.7
Flavor			
Roasted coffee essence	7.2	5.9	4.6
base bean	5.0	4.3	4.6
flora/winey	2.5	1.8	0.0
Green bean	0.0	0.0	0.0
Grainy/cereal	3.5	4.2	5.4
Burnt	1.2	1.6	2.3
Cardboard	0.0	0.0	0.0
Phenol/creosote	1.8	2.2	2.3
Sour	5.1	4.7	4.0
Bitter	3.3	3.3	3.6
Astringent	3.7	4.7	5.1

Appendix 1
Basic Data Analysis Methods

INTRODUCTION

The first step in the analysis of any set of data is to get a sense of the general location and dispersion (or spread) of the measurements. Both graphical and tabular methods are available, and both should be used. The basic data summary methods should be used before any formal statistical data analyses (e.g., confidence intervals, tests of hypotheses, model fitting) are performed. Examination of the graphs and tables may reveal features of the data that would be lost in the computation of test statistics and probabilities (p-values). In fact, features revealed in the graphs and tables may indicate that standard statistical analyses would be inappropriate for the data at hand.

Many standard statistical procedures assume that the data are normally distributed. The p-values used to assess the statistical significance of various results are computed using this assumption. If the data are not normally distributed, the p-values are inaccurate. The level of inaccuracy increases as the measurements exhibit greater departures from normality. There is also a less theoretical issue involved. Many standard statistical procedures rely on the sample mean:

$$\overline{X} = (\sum_{i=1}^{n} X_i)/n$$

and the sample standard deviation:

$$S = \sqrt{[\sum_{i=1}^{n} X_i^2 - (\sum_{i=1}^{n} X_i)^2/n]/(n-1)}$$

to summarize the measurements. The sample mean locates the center of the measurements. The sample standard deviation is a measure of the dispersion of the data about $\overline{X}$. For highly skewed or multi-modal data, $\overline{X}$ does not provide a good measure of the center of the data. $\overline{X}$ is sensitive to extreme values in the data and, therefore, overreacts when the measurements are highly skewed. Multi-modal behavior may indicate that the data arise from a mixture of several distributions, each with their own underlying mean value. The single value of $\overline{X}$ is meaningless in such situations. Further, since S measures the spread around $\overline{X}$, it too is a poor summary measure for skewed or multi-modal data. The single value of S implies that the data are dispersed symmetrically about $\overline{X}$ (not true for skewed data) and that only a single mean exists (not true for multi-modal data).

GRAPHICAL SUMMARIES

Consider the sweetness intensity data (measured on a 15 cm line scale) of 50 consecutive batches of a powdered soft drink presented in Table A-1. Both the histogram and the corresponding frequency distribution of these measurements suggest that the data are relatively symmetric (see Fig. A-1). This observation is confirmed by the box-and-whisker plot in Figure A-2.

A box-and-whisker plot is an informative graphical summary for medium to large data sets (i.e., 25 or more observations). The plot is easy to construct because it is based directly on the observed values and not on any complicated summary statistics. For instance, the "box" in Figure A-2 is bounded by the first (Q_1) and third (Q_3) quartiles of the data—that is, the points at which 25 percent of the data and 75 percent of the data fall below, respectively. The plus sign in the box is the

Table A-1.   Sweetness intensity ratings on 50 consecutively sampled batches of powdered soft drink

8.1	7.7	8.0	7.3	7.3	8.7	8.0	8.8	8.7	7.6	8.7
8.5	8.9	7.2	8.4	7.3	9.0	7.6	8.0	8.1	8.1	7.6
8.4	7.7	8.5	7.6	8.0	7.4	7.3	8.1	8.6	7.5	8.1
8.2	7.7	8.7	7.4	6.8	7.5	8.2	7.5	8.0	8.4	8.3
7.5	8.9	9.2	7.4	8.3	8.3					

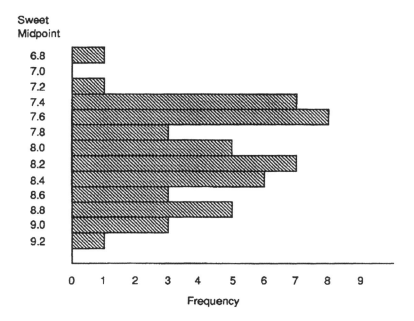

FIGURE A-1.   Histogram of the sweetness intensity data from Table A-1.

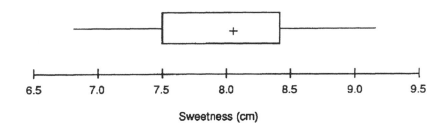

FIGURE A-2.   Box-and-whisker plot of the sweetness intensity data from Table A-1.

second quartile (or median, M); the point at which 50 percent of the data fall below. $Q_1 = 7.5$, $Q_3 = 8.4$, and M = 8.05 are presented with the summary statistics for the sweetness intensity data in Table A-2. The "whiskers" that extend beyond the box on the low and high sides are calculated as:

1. The maximum of L = $M-1.5(Q_3-Q_1)$ and the minimum data value. For the

Table A-2.   Summary statistics on the sweetness intensity data in Table A-1

Statistic	Value
Count (n)	50
Mean ($\overline{X}$)	8.02
Median (M)	8.05
Standard Deviation. (S)	0.56
Standard Error (SE)	0.08
Minimum	6.8
Maximum	9.2
First Quartile (Q1)	7.5
Third Quartile (Q3)	8.4

sweetness data, L = 8.05 - 1.5(8.4 - 7.5) = 6.7, and the minimum data value is 6.8 (from Table A-2), so the lower whisker of the box-plot extends downward from 7.5 to 6.8.

2. The minimum of U = $M + 1.5(Q_3 - Q_1)$ and the maximum data value. For the sweetness data, U = 8.05 + 1.5(8.4 - 7.5) = 9.4, and the maximum data value is 9.2 (from Table A-2), so the upper whisker of the box-plot extends upward from 8.4 to 9.2.

Data points that fall beyond the whiskers are plotted individually.

The two "valleys" in the histogram (at midpoints 7.8 and 8.6) in Figure A-1 hint at the possibility of multi-modal behavior. However, an additional graphical technique leads to the conclusion that this is not a serious concern. The normal probability plot in Figure A-3 displays the observed measurements (on the vertical axis) versus the normal deviate scores of n=50 measurements (on the horizontal axis). To construct a normal probability plot, the measurements are first ranked in increasing order—i = 1 to n. The ith ranked observation is then plotted versus $P = \Phi^{-1}[(3i-1)/(3n+1)]$, where Φ^{-1} is the inverse of the cumulative normal distribution function (see Table A-3). Many statistical analysis packages provide a programmed function to compute these values. If the observed data form a straight-line relationship with the normal scores, as is the case in Figure A-3, it can be concluded that the measurements arise from a normal distribution. A small number of observations that fall far from an otherwise straight line may be outliers in a set of normally distributed data. If the data form a curvilinear relationship with the normal scores, then they arise from a non-normal distribution (see D'Agostino, Belanger, and D'Agostino 1990).

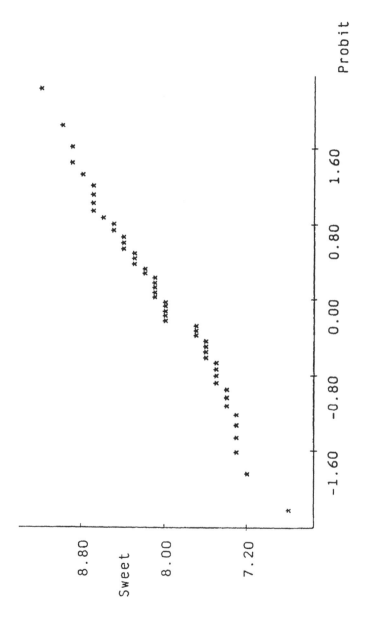

FIGURE A-3.   Normal probability plot of the sweetness intensity data from Table A-1.

Table A-3.   Sweetness data sorted in increasing order with normal distribution values used in the normal probability plot in Figure A-3

Rank(i)	$\Phi^{-1}[(3i-1)/(3n+1)]$	Sweetness
1	-2.21895	6.8
2	-1.83690	7.2
3	-1.61662	7.3
4	-1.45491	7.3
5	-1.32422	7.3
6	-1.21291	7.3
7	-1.11488	7.4
8	-1.02654	7.4
9	-0.94556	7.4
10	-0.87036	7.5
11	-0.79978	7.5
12	-0.73297	7.5
13	-0.66929	7.5
14	-0.60821	7.6
15	-0.54933	7.6
16	-0.49229	7.6
17	-0.43681	7.6
18	-0.38264	7.7
19	-0.32957	7.7
20	-0.27742	7.7
21	-0.22601	8.0
22	-0.17519	8.0
23	-0.12482	8.0
24	-0.07477	8.0
25	-0.02490	8.0
26	0.02490	8.1
27	0.07477	8.1
28	0.12482	8.1
29	0.17519	8.1
30	0.22601	8.1
31	0.27742	8.2
32	0.32957	8.2
33	0.38264	8.3
34	0.43681	8.3
35	0.49229	8.3
36	0.54933	8.4
37	0.60821	8.4
38	0.66929	8.4
39	0.73297	8.5
40	0.79978	8.5
41	0.87036	8.6
42	0.94556	8.7
43	1.02654	8.7
44	1.11488	8.7
45	1.21291	8.7
46	1.32422	8.8
47	1.45491	8.9
48	1.61662	8.9
49	1.83690	9.0
50	2.21895	9.2

SUMMARY STATISTICS

Based on the earlier graphical inspection of the data, the standard descriptive statistics are reliable summaries of the sweetness intensity measurements. Table A-2 contains these statistics, as well as those required to generate the box-and-whisker plot in Figure A-2. The estimate of the standard error of the mean (SE) is also included. The standard error of the mean is the standard deviation of the distribution of $\overline{X}$'s of a specified sample size—n. If the raw data have a standard deviation σ, the standard error of the mean is $\sigma/\sqrt{n}$. Note that as n increases, $\sigma/\sqrt{n}$ decreases, so for larger sample sizes, $\overline{X}$ is more likely to take on a value close to its true average—μ. The standard error of the mean is estimated by SE = $S/\sqrt{n}$

CONFIDENCE INTERVALS

The values of $\overline{X}$ and SE in Table A-2 can be used to decide if $\overline{X}$ is a sufficiently precise estimate of μ to meet the needs of an investigation. This is done by using confidence intervals. A confidence interval on a mean is a range of values within which the true value of μ lies with a known probability. Confidence intervals on the mean are calculated as:

$$\overline{X} \pm t_{\alpha/2, n-1}SE$$

where $t_{\alpha/2,n-1}$, is the critical value from Student's t distribution. The quantity α measures the level of confidence. For instance, if $\alpha=0.05$, the confidence interval is a $100(1-\alpha)\% = 95\%$ confidence interval. The quantity (n-1) is a parameter of the t distribution called degrees-of-freedom. Degrees-of-freedom measure how much information is available to estimate the variability in a set of data. The value of t depends on the value of α and the number of degrees-of-freedom (n-1). Critical values of t are presented in Table A-8 at the end of the appendices. For the sweetness intensity data, a 95-percent confidence interval on the mean is:

$$8.02 + 2.01\,(0.079)$$
$$\text{or}$$
$$(7.86, 8.18)$$

So it can be concluded that with 95-percent confidence the true value of the average sweetness intensity of the powdered soft drinks lies somewhere between 7.86 cm and 8.18 cm. If this range of values is not sufficiently narrow, more observations need to be collected to further decrease SE and, as a result, the width of the interval.

Appendix 2

Statistical Hypothesis Testing

INTRODUCTION

Statistical hypothesis testing is a decision-making technique that quantifies the risks associated with various decisions and, thereby, increases the comfort level in every decision made. The first step in a statistical hypothesis test is to state two mutually exclusive hypotheses about the true state of a system. The first of these hypotheses, the null hypothesis, is the condition that is assumed to exist prior to running a study. The value specified in the null hypothesis is used to calculate the test statistic (and resulting p-value) in the hypothesis test. The second of these hypotheses, the alternative hypothesis, is developed based on the prior interest of the investigator.

The alternative hypothesis is generally of greater interest to a researcher because, when true, it indicates that some action is called for. For example, if a company is replacing one of the raw ingredients in its current product with a less expensive one, the primary concern is that the product made with the less expensive ingredient cannot be distinguished from the current product. The null hypothesis and the alternative hypothesis for this investigation are:

$$H_o : \mu_{\text{current}} = \mu_{\text{less expensive}}$$

vs.

$$H_a : \mu_{\text{current}} \neq \mu_{\text{less expensive}}$$

where μ is the mean value of any critical product characteristic of product i. Both the null and the alternative hypotheses must be specified before the test is conducted. Results are biased in favor of rejecting the null hypothesis too often if the alternative hypothesis is formulated after reviewing the data.

TYPE I AND TYPE II ERRORS

In testing statistical hypotheses, some conclusion is drawn. The conclusion may be correct or incorrect. There are two ways in which an incorrect conclusion may be drawn. First, a researcher may conclude that the null hypothesis is false, when, in fact, it is true (e.g., conclude that a difference exists when it does not). Such an error is called a Type I error. Second, a researcher may conclude that the null hypothesis is true or, more correctly, that the null hypothesis cannot be rejected, when, in fact, it is false (e.g., failing to detect a difference that exists). Such an error is called a Type II error. The probability of making Type I and Type II errors are specified before the investigation is conducted. The probability of making a Type I error is equal to α. The probability of making a Type II error is equal to β. α is called the producer's risk because it is the probability that an acceptable product (i.e., in H_o) is incorrectly determined to be unacceptable and, therefore, not sold. β is called the consumer's risk because it is the probability that unacceptable product (i.e., in H_a) fails to be identified and, as a result, is accepted from the producer.

Trial by jury is a useful analogy to hypothesis testing. The null hypothesis is that the defendant is innocent. The alternative hypothesis is that the defendant is not innocent (i.e., guilty). Convicting an innocent person is a Type I error. Since Type I errors are undesirable (particularly for the innocent person), convictions are only handed down when a person has been proven guilty beyond a reasonable doubt. In hypothesis testing, "reasonable doubt" is quantified as α. Failing to convict a guilty person is a Type II error. The probability of letting a guilty person go free is β. Both α and β should be kept small. This is done by presenting a sufficient amount of evidence at the trial—in general, the more the better. In hypothesis testing, the sampled units of product are the evidence. To contain the risks of drawing incorrect conclusions within known limits (i.e., α and β), a sufficiently large number of units need to be sampled (see Kraemer and Thiemann 1987).

EXAMPLE OF A STATISTICAL HYPOTHESIS TEST

Consider the sweetness data in Table A-1. Suppose the investigator needed to have a high level of confidence that the mean sweetness rating of the product was greater than 7.5 cm. The null hypothesis is $H:\mu \leq 7.5$. The alternative hypothesis is $H_a:\mu > 7.5$. The test statistic is:

$$t = \frac{\overline{X} - \mu_o}{SE}$$

where $\overline{X}$ and SE are presented in Table A-2 and μ_o is the assumed value of the mean sweetness from the null hypothesis (or, as in this example, the "worst case" situation from H_o). Plugging in the appropriate values yields:

$$t = \frac{8.02 - 7.5}{0.079} = 6.58$$

The statistic t, above, follows a Student's t distribution with n-1 = 50-1 = 49 degrees of freedom. Choosing α = 0.01, the critical value of t from Table A-8 is found to be $t_{0.01,49}$=2.41. Since the calculated value t = 6.58 exceeds the critical value from the table, the null hypothesis is rejected at the 1-percent level of significance. In fact, the probability of obtaining a value of $t_{49} \geq 6.58$ by chance alone (i.e., if H_o is true) is less than 1 in 10,000. The investigator can safely decide to reject the null hypothesis that the mean sweetness rating of the powdered soft drinks averages no more than 7.5 cm in favor of the alternative hypothesis that the sweetness intensity is greater than 7.5 cm.

This example is a very sensitive hypothesis test. Given the sample size, n = 50, and the small level of variability, SE = 0.079, the investigator has less than β = 0.0001 chance of failing to detect the case where the average sweetness rating is 8.0 cm or more.

Appendix 3

The Statistical Design of Sensory Panels

INTRODUCTION

The basic goal of the statistical analysis of a designed study is to obtain an accurate and precise estimate of experimental error. All tests of hypotheses and confidence statements are based on this. Experimental error is the unexplainable, natural variability of the products or samples being studied. Typically, experimental error is expressed quantitatively as the sample standard deviation, S, or as the sample variance, S^2. In order to obtain a good estimate of experimental error, all sources of "non-product" variability that are known to exist before a study is run should be compensated for.

One important source of variability in sensory analysis that is known to exist in every study is panelist-to-panelist variability. For a variety of reasons, panelists may use different parts of the rating scale to express their perceptions of the intensities of a product's attributes. Sensory panels need to be designed to account for this behavior so as to obtain the most sensitive comparisons among the samples being evaluated. The statistical technique known as "blocking" accomplishes this.

Sensory panels are composed of a large number of evaluations. The evaluations are grouped together according to panelists (i.e., the blocking factor), in recognition of the fact that they may use different parts of the rating scale to express their perceptions. The samples must be independently applied at each evaluation. This is accomplished through such techniques as randomized orders of presentation, sequential monadic presentations, and wash-out periods of sufficient duration to allow the panelist to return to some baseline level (constant for all evaluations).

RANDOMIZED (COMPLETE) BLOCK DESIGNS

If the number of samples is small enough so that sensory fatigue is not a concern, a randomized (complete) block design is appropriate. Each panelist evaluates all of the samples (thus the term "complete block"). A randomized block design is effective when the investigator is confident that the panelists are consistent in rating the samples. If, during training for example, the investigator has doubts as to whether all of the panelists are rating the samples in the same way, then a more complicated split-plot design should be used (see Meilgaard, Civille, and Carr 1991).

Independently replicated samples of the test products are presented to the panelists in a randomized order (using a separate randomization for each panelist). The data obtained from the panelists' evaluations can be arranged in a two-way table, as in Table A-4. The data from a randomized block design are analyzed by analysis of variance (ANOVA). The form of the ANOVA table appropriate for a randomized block design is presented in Table A-5. The null hypothesis is that the

Table A-4.   Data table for a randomized (complete) block design

Blocks (Judges)	Samples					
	1	2	.	.	.	t
1	X_{11}	X_{12}	.	.	.	X_{1t}
2	X_{21}	X_{22}	.	.	.	X_{2t}
.	.	.				.
.	.	.				.
.	.	.				.
b	X_{b1}	X_{b2}	.	.	.	X_{bt}

Table A-5.   ANOVA table for a randomized (complete) block design

Source of Variability	Degrees of Freedom	Sum of Squares	Mean Square	F
Total	bt-1	SS_T		
Blocks (Judges)	b-1	SS_J		
Samples	$df_S = t-1$	SS_S	$MS_S = SS_S/df_S$	MS_S/MS_E
Error	$df_E = (b-1)(t-1)$	SS_E	$MS_E = SS_E/df_E$	

mean ratings for all of the samples are equal (H_o: $\mu_i = \mu_j$ for all samples i and j) versus the alternative hypothesis that the mean ratings of at least two of the samples are different (H_a: $\mu_i \neq \mu_j$ for some pair of distinct samples i and j). For t samples, each evaluated by b panelists, if the value of the F-statistic calculated in Table A-5 exceeds the critical value of an F with (t-1) and (b-1)(t-1) degrees of freedom (Table A-9 at the end of the appendices), the null hypothesis is rejected in favor of the alternative hypothesis.

A significant F-statistic in Table A-5 raises the question of which of the samples differ significantly. This question is answered by another statistical technique called a multiple comparison procedure. To determine which samples have significantly different mean ratings, a Fisher's LSD for randomized (complete) block designs is used, where:

$$LSD = t_{\alpha/2,\, df_E} \sqrt{2MS_E/b}$$

where b is the number of blocks (typically judges) in the study, $t_{\alpha/2, df_E}$ is the upper $\alpha/2$ critical value of the Student's t distribution with df_E degrees of freedom, and MS_E is the mean square for error from the ANOVA table. Any two sample means that differ by more than LSD are significantly different at the α level.

BALANCED INCOMPLETE BLOCK DESIGNS

If the number of samples in a study is greater than what can be evaluated before sensory fatigue sets in, a balanced incomplete block design (BIB) should be used to assign samples to each panelist. In BIB designs, the panelists evaluate only a portion, k, of the total number of samples, t (k < t). The specific set of k samples that a panelist evaluates is selected so that, in a single repetition of a BIB design, every sample is evaluated an equal number of times (denoted by r) and all pairs of samples are evaluated together an equal number of times (denoted by λ). In a BIB design, each sample mean is estimated with equal precision, and all pair-wise comparisons between two sample means are equally sensitive. The number of blocks required to complete a single repetition of a BIB design is denoted by b. Table A-6 illustrates a typical BIB layout. A list of BIB designs, such as the one presented by Cochran and Cox (1957), can be helpful in selecting a specific design for a study.

In order to obtain a sufficiently large number of total replications, the entire BIB design (b blocks) may have to be repeated several times. The number of repeats or repetitions of the fundamental design is denoted by p. The total number of blocks is then pb. This yields a total of pr evaluations of every sample and a total of $p\lambda$ direct comparisons for any two samples occurring in the BIB design.

If the number of blocks in the BIB design is relatively small (four or five, say), it may be possible to have a small number of panelists (p in all) return several times

Table A-6.   Sample assignments to blocks for a balanced incomplete block
Design (t=7, k=3, b=7, r=3, λ=1, p=1)

Sample Block	1	2	3	4	5	6	7
1	x	x		x			
2		x	x		x		
3			x	x		x	
4				x	x		x
5	x				x	x	
6		x				x	x
7	x		x				x

until each panelist has completed an entire repetition of the design. The order of
presentation of the blocks should be randomized separately for each panelist, as
should be the order of presentation of the samples within each block. Alternatively,
for BIB designs with large numbers of blocks, the normal practice is to call upon
a large number of panelists (pb in all) and to have each evaluate a single block of
samples. The block of samples that a particular panelist receives should be
assigned at random. The order of presentation of the samples within each block
should again be randomized in all cases.

ANOVA is used to analyze BIB data (see Table A-7). As in the case of a
randomized (complete) block design, the total variability is partitioned into the
separate effects of blocks, samples, and error. However, the formulas used to
calculate the sum of squares in a BIB analysis are more complicated than for a
randomized (complete) block analysis (see Kirk 1968). Regardless of the approach
used to run the BIB design, the ANOVA table presented in Table A-7 partitions the
sources of variability so that clean estimates of the sample effects and uninflated

Table A-7.   ANOVA table for a balanced incomplete block designs

Source of Variability	Degrees of Freedom	Sum of Square	Mean Square	F
Total	tpr-1	SS_T		
Blocks	pb-1	SS_B		
Samples (adj. for Blocks)	$df_S = t-1$	SS_S	$MS_S = SS_S/df_S$	MS_S/MS_E
Error	$df_E = tpr-t-pb+1$	SS_E	$MS_E = SS_E/df_E$	

estimates of experimental error (even in the presence of panelist-to-panelist variability) are obtained.

If the F-statistic in Table A-7 exceeds the critical value of an F with the corresponding degrees of freedom, the null hypothesis assumption of equivalent mean ratings among the samples is rejected. Fisher's LSD for BIB designs has the form:

$$LSD = t_{\alpha/2,\ df_E}\sqrt{2MS_E/pr}\ \sqrt{[k\,(t-1)]/[\,(k-1)\,t]}$$

where t is the total number of samples, k is the number of samples evaluated by each panelist during a single session, r is the number of times each sample is evaluated in the fundamental design (i.e., in one repetition of b blocks), and p is the number of times the fundamental design is repeated. MS_E and $t_{\alpha/2,df_E}$ are as defined before.

Appendix 4

Multivariate Methods

INTRODUCTION

Multivariate statistical methods are used to study groups of responses (i.e., multiple variables) that are simultaneously collected on each unit in a sample. The methods take into consideration the existence of groups of correlated variables—that is, groups of variables whose values increase and decrease together (either in direct or inverse relationship to one and other). Multivariate methods are particularly useful for analyzing consumer test data and descriptive data where several consumer acceptability and descriptive intensity ratings are taken on each sample evaluated. Multivariate methods can provide concise summaries of the total variability, using fewer measurements than originally collected, or they may identify subsets of respondents who display different patterns of responses to the samples than other respondents in the study.

MULTIVARIATE SUMMARY STATISTICS

The summary analysis of multivariate data is a direct extension of univariate techniques. The center of the multivariate population is stated in the mean vector:

$$\underline{\mu} = \begin{bmatrix} \mu_1 \\ \mu_2 \\ \cdot \\ \cdot \\ \cdot \\ \mu_p \end{bmatrix}$$

where each component of the vector is the univariate mean for each response. The estimator of μ is the vector, $\bar{X}$, of individual sample means, $\bar{X}$'s. The dispersion of a multivariate population is summarized in the variance-covariance matrix, Σ. The estimator of Σ is the sample covariance matrix, S, of sample variances and sample covariances, where:

$$S_i^2 = \sum_{h=1}^{n} (X_{hi} - \bar{X_i})^2/(n-1)$$

form the main diagonal of the matrix, and:

$$S_{ij} = \sum_{h=1}^{n} (X_{hi} - \bar{X_i})(X_{hj} - \bar{X_j})/(n-1)$$

are the off-diagonal elements. $\bar{X}$ and S are used in multivariate tests of hypotheses and confidence intervals, just as $\bar{X}$ and S are used in univariate situations (see Morrison 1976).

PRINCIPAL COMPONENTS/FACTOR ANALYSIS

A multivariate statistical technique that takes advantage of the covarying structure of groups of variables is principal components/factor analysis (PCA/FA). Although PCA and FA have different underlying theories (see Smith 1988), their application in sensory evaluation is similar. Therefore, they are treated together here. Computer programs that extract the principal components from a set of multivariate data are widely available, so computational details are not included in the following discussion.

PCA/FA are dimension-reducing techniques that allow the variability in a large number of product characteristics to be displayed in a relatively small number of dimensions. Each of the new dimensions, or principal components, is selected to be uncorrelated with any others and such that each successive principal component explains as much of the remaining unexplained variability as possible. It is often possible to explain more than 80 to 90 percent of the total variability in as many as 25 to 30 product characteristics, with as few as two or three principal components.

A plot of the principal component scores for a set of products can reveal groupings and polarizations of the samples that would not be as readily apparent in an examination of the larger number of original variables (see Fig. A-4). Principal component scores might also be used to define a range of acceptable product variability that would be less apparent or less sensitively defined if each of the original variables was considered individually. This could be done by simply

First and Second Principal Components

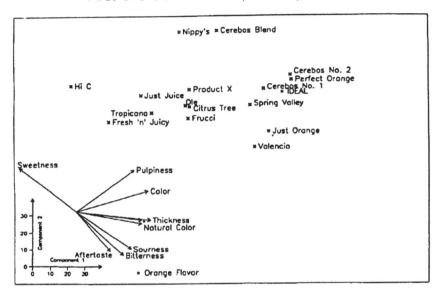

FIGURE A-4.   Plot of the first two principal component scores for a set of 16 orange juice products, showing both the orientation of the original attributes and the distribution of the products. (Copyright ASTM. Reprinted with permission.)

plotting the products overall acceptability ratings versus each of the principal component scores and determining if a range of values that relates to meaningful acceptability limits exists (see Fig. A-5).

Each principal component, y_i, is a linear combination of the original observations, x_j, as in:

$$y_i = a_{i1}x_1 + a_{i2}x_2 + \ldots + a_{ip}x_p$$

The coefficients, a_{ij}, lie between -1 and +1 and measure the importance of the original variables on each principal component. A coefficient close to -1 or +1 indicates that the corresponding variable has a large influence on the value of the principal component; values close to zero indicate that the corresponding variable has little influence on the principal component. In general, the original variables tend to segregate themselves into nonoverlapping groups where each group is associated predominantly with a specific principal component. These groupings help in interpreting the principal components and in ascertaining which of the original variables combine to yield some net effect on the product.

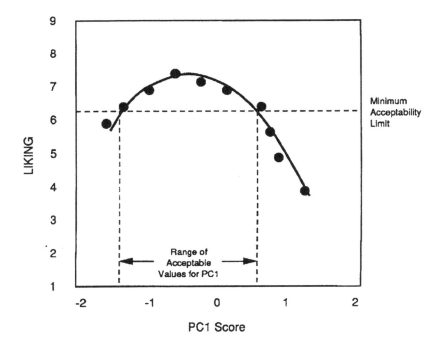

FIGURE A-5.   Plot of the relationship between liking and the first principal component scores for a set of ten products.

The number of original variables studied should not be reduced based on the results of a PCA/FA analysis. As can be seen in the equation for y_i, each of the original variables is included in the computation of each principal component. Retaining only a small group of "representative" variables on a sensory ballot ignores the multivariate nature of the effects of the original variables and, if done, may lead to misleading results in future evaluations.

CLUSTER ANALYSIS

Another multivariate technique that those working in the area of QC/sensory should be aware of is cluster (or segmentation) analysis. The details of the computations involved in cluster analysis go beyond the scope of the present discussion. In fact, the discussion of the relative merits of various clustering methods continues (see Jacobsen and Gunderson 1986). The goal of cluster analysis is to identify homogeneous subgroups (i.e., clusters) of individuals (either respondents or products) based on their degree of similarity in certain characteristics. Some work on clustering sensory attributes has also been done (see Powers, Godwin, and Bargmann 1977). Although the statistical propriety of such an analysis is questionable (because the attributes on a sensory ballot are not a random sample from any population), the method still provides an informative way to study the degree of similarity and groupings of attributes used to measure product quality.

An application of cluster analysis particularly important in QC/sensory is that of identifying segments of consumers that have different patterns of liking across products. While some consumers may like a highly toasted note in a cereal product, others may find this objectionable. The net effect of merging these two groups would be to obscure their individual trends and, thus, to fail to identify a meaningful criteria of product quality in the eyes of consumers. In some ways, failing to recognize clusters of respondents is equivalent to computing the mean of a multi-modal set of data. The "middle" of such a set may well be the valley between two groups of data where no individuals exist. Developing a product to suit this "average" respondent may, in fact, please no one.

The application of cluster analysis to identify segments of respondents that have different liking patterns proceeds through the following five steps:

1. Conduct an acceptability test on a variety of products known to span the typical ranges of important product characteristics.
2. Submit each respondent's overall acceptability ratings on each sample of product to a cluster analysis program and obtain a summary of the clusters, such as the tree diagram in Figure A-6.
3. Separately for each cluster, perform correlation and regression analyses to determine how the perceived product attributes relate to overall acceptability.
4. Determine the acceptability limits of each product attribute for each segment of respondents, as done in the comprehensive approach presented in Chapter 3.
5. If conflicting acceptability limits are obtained, examine the demographic information on each segment of respondents to determine which segment is more important to the product's success (e.g., heavy purchasers, target age group, sex).

Cluster analysis can also be used to identify groups of production samples according to the intensity of the descriptive sensory attributes or other measures of product quality. The ratings of each product would be submitted to a cluster

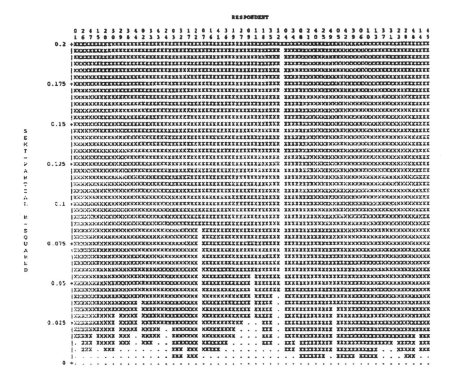

FIGURE A-6.   A tree diagram used in cluster analysis to identify groups of respondents with similar patterns of liking across products.

analysis program to obtain a summary of the clusters of products (similar to the tree diagram in Figure A-6). The average ratings of the product attributes could then be computed within each cluster to determine how products from different clusters differentiate themselves. Lastly, production log sheets could be examined to determine if identifiable causes such as plant, day, shift, and line play a role in determining cluster membership.

Appendix 5

Statistical Quality Control

INTRODUCTION

There will always be some variability in the output of a production process. This variability can be broken down into two types—assignable causes and common (or random) causes. Assignable causes of variability are those over which some control can be exerted, such as sources of raw materials, process settings, or the level of training of line operators. The common causes of variability are the more complex sources inherent to the process over which no direct control is available. The overall effects of common causes are generally small.

A production process is said to be "in control," statistically, when the product's attributes are varying around their target levels in a manner influenced by only common causes. If variability due to one or more assignable causes results in excessive product variability and/or a shift away from the target levels, the process is said to be "out of control," statistically.

CONTROL CHARTS

Control charts are simple graphical devices for monitoring the output of a process to determine if it is in control. Statistical criteria, similar to hypothesis tests, allow users of control charts to distinguish between random variability and assignable causes. Thus, the number of unneeded process adjustments is minimized and, simultaneously, early warnings of out-of-control conditions are obtained. The warnings from control charts are often issued before the process is producing product that is out-of-specification and, therefore, must be scrapped.

X̄-Charts

The most common control chart is the X̄-chart, which monitors the average value of a product variable. To use an X̄-chart, a small, fixed number of samples (typically, three to five) needs to be drawn at regular intervals (e.g., within a batch) during production. The critical measures of quality are taken on each of the units sampled. For each variable measured, the sample mean of the small group is computed and plotted on the X̄-chart. The mean of each new group is plotted immediately following the mean of the previous group. If the process is in control, then the X̄'s should all vary about their target, μ, within a narrow range defined by the common causes of variability. The range is defined by the historical variability of the process and is summarized by the standard deviation σ. At least 25 groups of samples (each determined to be in control) should be used to compute μ and σ (see Fig. A-7).

Nelson (1984) proposed eight criteria for detecting assignable causes of variability in an X̄-chart. Failing to meet any one of these criterion would indicate an out-of-control condition existed. The criteria are:

1. One point beyond $\mu \pm 3\sigma/\sqrt{n}$.
2. Nine points in a row, all on one side of μ.

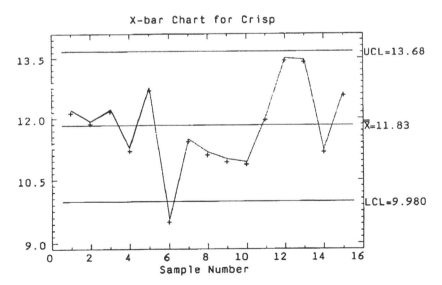

FIGURE A-7.   An X̄-chart of the average crispness of a product showing the latest 15 production batches.

3. Six points in a row, all increasing or all decreasing.
4. Fourteen points in a row, alternating up and down.
5. Two out of three points in a row beyond $\mu \pm 2\sigma/\sqrt{n}$.
6. Four out of five points in a row beyond $\mu \pm 1\sigma/\sqrt{n}$ (on one side of μ).
7. Fifteen points in a row within $\mu \pm 1\sigma/\sqrt{n}$ above and below μ.
8. Eight points in a row beyond $\mu \pm 1\sigma/\sqrt{n}$ above and below μ.

Nelson's first criterion is the standard "action limit" of any $\overline{X}$ falling outside of $\mu \pm 3\sigma/\sqrt{n}$. The $\pm 3\sigma/\sqrt{n}$ limits are also called the upper control limit (UCL) and lower control limit (LCL), respectively. When an individual $\overline{X}$ exceeds the action limit, steps are taken to bring the process back into control. The balance of the criteria go beyond the standard "warning limits" of any $\overline{X}$ falling beyond $\mu \pm 2\sigma/\sqrt{n}$. An individual $\overline{X}$ that exceeds the warning limits triggers a search for the cause of the change, but no adjustments to the process are made.

R-Charts

The sample mean, $\overline{X}$, measures how close the product is to its target level, μ, "on the average." The range, R (= maximum - minimum values in the sample), is a simple measure of the dispersion of the individual readings of product quality. An extremely large or extremely small value of R is another indication of an out-of-control situation. To monitor the variability of the product, an R-chart is constructed by plotting the ranges of the periodic samples (i.e., the same samples used to compute the sample mean in the $\overline{X}$-chart) during production (see Fig. A-8).

Statistical criteria exist for determining if a process is out of control based on the value of the range. The criteria are equivalent to the $\pm 2\sigma/\sqrt{n}$ and $\pm 3\sigma/\sqrt{n}$ limits used in the $\overline{X}$-chart. The warning and action limits (i.e., the 95-percent and 99.9-percent confidence level limits, respectively) are computed by multiplying the average range $\overline{R}$ by the appropriate value in Table A-10, at the end of the appendices. An individual value of R that exceeds the warning limits has less than a 5-percent chance of occurring by chance alone. Similarly, an individual value of R that exceeds the action limits has less than one chance in 1,000 of occurring while the process is in control. Either situation indicates an increased potential for assignable causes to be affecting the process. However, adjustments are made to the process only when an action limit has been exceeded.

I-Charts

If only a single unit of product is collected at each of the periodic samplings, an I-chart can be constructed in the same way as an $\overline{X}$-chart. The standard deviation of the individual values can be calculated, and warning limits, action limits, and

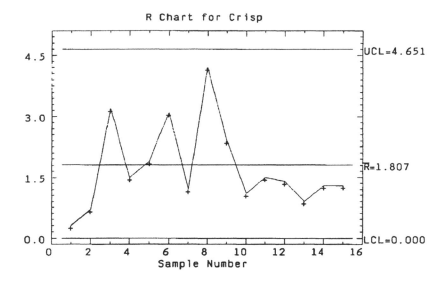

FIGURE A-8.   An R-chart of the within-batch range of crispness ratings of a product, showing the latest 15 production batches.

Nelson's criteria can be applied (see Fig. A-9). It is not possible to construct an R-chart when only a single unit is collected from each batch.

When sensory panel data are used in a statistical process control program, it is important to distinguish between the panel mean for an individual sample of product and the mean of a small group of production samples. If only a single sample of product is collected during each of the periodic QC samplings, the sensory panel yields only one piece of raw data (i.e., the panel mean) about the state of the process at that point in time. This is true, regardless of the number of individuals on the panel. The individual data values (i.e., the panel means) from successive QC samplings are plotted on an I-chart. If several samples of product are collected from each production batch, the samples should be presented to the panel according to an experimental design (e.g., a randomized complete block design) that will allow a panel mean to be calculated for each of the samples. The panel means of the group of samples are then used as the raw data to compute $\overline{X}$ and R that summarize the state of the process for that batch. If only three samples are collected within a batch, then only three pieces of raw data (i.e., the three panel means) are available to compute $\overline{X}$ and R, regardless of the number of panelists. The $\overline{X}$'s and R's from successive batches are plotted on $\overline{X}$-charts and R-charts.

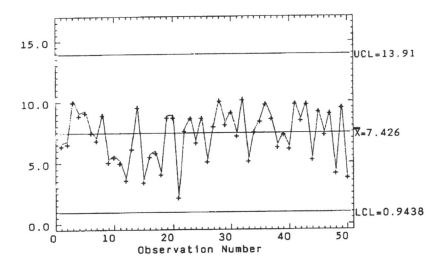

FIGURE A-9.   An I-chart of the firmness ratings of a product, based on a single sample collected from each of the last 50 production batches.

STATISTICAL CONTROL VERSUS SPECIFICATIONS

Thus far, the discussion has considered only statistical process control. No mention has been made of external specification limits (or tolerances). There is no inherent connection between external specifications and the statistical action and warning limits that are based on historical process variability. It should not be assumed that a production process that is statistically "in control" is producing "acceptable" (i.e., in spec) product. If the specification limits are narrow, this may not be true. Specifically, if the specification limits are less than 6σ, a reasonably large proportion of "in control" product may fall out-of-specification. When this happens, either the specifications need to be examined to determine if they can be widened or the process needs to be modified to reduce the common causes of variability (see Wetherill 1977).

Hopefully, the upper and lower control limits (e.g., the $\pm 3\sigma$ "action limits") will be sufficiently narrow to fit within the specification limits. Ideally, the control limits will be so narrow that the target level of the process can be set off center in the specification range, based on production costs or other criteria (see Fig. A-10).

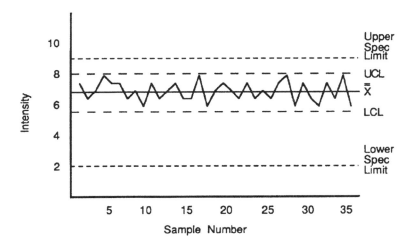

FIGURE A-10.   An $\overline{X}$-chart, including both control limits and specification limits, showing that the process does not have to operate at the center of the specification range.

PANEL MAINTENANCE

Two sources of variability have been discussed thus far—common causes and assignable causes. A third source of variability that is particularly important to sensory data is measurement variability. The instruments of sensory evaluation are panelists, who are known to be sensitive to a variety of factors that can influence their evaluations. In order to serve as a useful analytical measure of product quality, the output of a sensory panel must be sufficiently precise to be relied on. Data from a sensory panel should be held up to the same quality standards as any other analytical data. An analytical method is typically regarded as being "pretty good" if it has a relative standard deviation (RSD=S/$\overline{X}$) of 5 percent or less. Repeated evaluations of the same, or standard, samples should be performed periodically to insure that the panel's mean ratings are sufficiently precise.

There are two components of panel variability—within-session variability and between-session variability. Within-session variability results from the panelist-to-panelist differences. A measure of within-session variability is:

$$RSD_w = SE_w/\overline{X}_w$$

where SE_w is the sample standard error of the mean from a given session and $\overline{X}_w$ is the sample mean attribute rating from that session. RSD_w can be reduced by increasing the number of panelists who participate in the sessions (assuming they are all well trained and calibrated). Between-session variability arises from changes in the testing environment (odors, lighting, etc.), general shifts in the calibration of the panelists (motivation, fatigue, etc.), and so forth. Good analytical test controls must be exercised to keep these sources of variability to a minimum. Between session variability is measured by:

$$RSD_b = S_{\bar{X}}/\bar{\bar{X}}$$

$$\text{where } S_{\bar{X}} = \sqrt{\sum_{i=1}^{P} (\bar{X}_i - \bar{\bar{X}})^2/(p-1)},$$

X_i is the sample mean from session i (i = 1 to p sessions), and $\bar{\bar{X}}$ is the grand sample mean (i.e., the arithmetic average of the individual session means) based on repeated evaluations of the same, or standard, sample. When necessary, RSD_w and RSD_b can be compared to decide which source of variability needs attention.

Table A-8.   Two-tailed critical values for Student's t distribution

DF	0.10	0.05	0.01
1	6.31	12.70	63.65
2	2.92	4.30	9.92
3	2.35	3.18	5.84
4	2.13	2.77	4.60
5	2.01	2.57	4.03
6	1.94	2.44	3.70
7	1.89	2.36	3.49
8	1.86	2.30	3.35
9	1.83	2.26	3.25
10	1.81	2.22	3.16
11	1.79	2.20	3.10
12	1.78	2.17	3.05
13	1.77	2.16	3.01
14	1.76	2.14	2.97
15	1.75	2.13	2.94
16	1.74	2.12	2.92
17	1.74	2.11	2.89
18	1.73	2.10	2.87
19	1.72	2.09	2.86
20	1.72	2.08	2.84
21	1.72	2.08	2.83
22	1.71	2.07	2.81
23	1.71	2.06	2.80
24	1.71	2.06	2.79
25	1.70	2.06	2.78
26	1.70	2.05	2.77
27	1.70	2.05	2.77
28	1.70	2.04	2.76
29	1.69	2.04	2.75
30	1.69	2.04	2.75
40	1.68	2.02	2.70
60	1.67	2.00	2.66
120	1.65	1.98	2.61
Normal Dist.	1.64	1.96	2.57

Table A-9.   Upper 5 percent critical values for the F distribution with df₁ degrees of freedom in the numerator and df₂ degrees of freedom in the denominator.

df₂	1	2	3	4	5	6	7	8	9	10	20	30	40	50	100	∞
1	161	200	216	225	230	234	237	239	241	242	248	250	251	252	253	254
2	18.51	19.00	19.16	19.25	19.30	19.33	19.36	19.37	19.38	19.39	19.44	19.46	19.47	19.47	19.49	19.50
3	10.13	9.55	9.28	9.12	9.01	8.94	8.88	8.84	8.81	8.78	8.66	8.62	8.60	8.58	8.56	8.53
4	7.71	6.94	6.59	6.39	6.26	6.16	6.09	6.04	6.00	5.96	5.80	5.74	5.71	5.70	5.66	5.63
5	6.61	5.79	5.41	5.19	5.05	4.95	4.88	4.82	4.78	4.74	4.56	4.50	4.46	4.44	4.40	4.36
6	5.99	5.14	4.76	4.53	4.39	4.28	4.21	4.15	4.10	4.06	3.87	3.81	3.77	3.75	3.71	3.67
7	5.59	4.74	4.35	4.12	3.97	3.87	3.79	3.73	3.68	3.63	3.44	3.38	3.34	3.32	3.28	3.23
8	5.32	4.46	4.07	3.84	3.69	3.58	3.50	3.44	3.39	3.34	3.15	3.08	3.05	3.03	2.98	2.93
9	5.12	4.26	3.86	3.63	3.48	3.37	3.29	3.23	3.18	3.13	2.93	2.86	2.82	2.80	2.76	2.71
10	4.96	4.10	3.71	3.48	3.33	3.22	3.14	3.07	3.02	2.97	2.77	2.70	2.67	2.64	2.59	2.54
20	4.35	3.49	3.10	2.87	2.71	2.60	2.52	2.45	2.40	2.35	2.12	2.04	1.99	1.96	1.90	1.84
30	4.17	3.32	2.92	2.69	2.53	2.42	2.34	2.27	2.21	2.16	1.93	1.84	1.79	1.76	1.69	1.62
40	4.08	3.23	2.84	2.61	2.45	2.34	2.25	2.18	2.12	2.07	1.84	1.74	1.69	1.66	1.59	1.51
50	4.03	3.18	2.79	2.56	2.40	2.29	2.20	2.13	2.07	2.02	1.78	1.69	1.63	1.60	1.52	1.44
100	3.94	3.09	2.70	2.46	2.30	2.19	2.10	2.03	1.97	1.92	1.68	1.57	1.51	1.48	1.39	1.28
∞	3.84	2.99	2.60	2.37	2.21	2.09	2.01	1.94	1.88	1.83	1.57	1.46	1.40	1.35	1.24	1.00

Table A-10.   Values for computing the lower action limit (LAL), the lower warning limit (LWL), upper warning limit (UWL), and the upper action limit (UAL) for R-charts

Limits are computed by multiplying the average range, $\overline{R}$, by the values given for the corresponding sample size.

Sample Size	LAL	LWL	UWL	UAL
2	0.00	0.04	2.81	4.12
3	0.04	0.18	2.17	2.99
4	0.10	0.29	1.93	2.58
5	0.16	0.37	1.81	2.36
6	0.21	0.42	1.72	2.22
7	0.26	0.46	1.66	2.12
8	0.29	0.50	1.62	2.04
9	0.32	0.52	1.58	1.99
10	0.35	0.54	1.56	1.94
11	0.38	0.56	1.53	1.90
12	0.40	0.58	1.51	1.87

References

American Oil Chemist Society. 1989. AOCS recommended practice Cg 2-83. Champaign, IL: AOCS.

American Society for Testing and Materials (ASTM). 1991. *A guide for the sensory evaluation function within a manufacturing quality assurance/quality control program*, ed. J. Yantis. Philadelphia: ASTM (in preparation).

American Society for Testing and Materials (ASTM). 1986. *Physical requirement guidelines for sensory evaluation laboratories*, ed. J. Eggert and K. Zook. Philadelphia: ASTM.

Amerine, M. A., R. M. Pangborn, and E. B. Roessler. 1965. *Principles of Sensory Evaluation of Food*. New York: Academic Press.

Amerine, M. A., E. B. Roessler, and F. Filipello. 1959. Modern sensory methods of evaluating wines. *Hilgardia* 28:477-567.

Aust, L. B., M. C. Gacula, S. A. Beard, and R. W. Washam II. 1985. Degree of difference test method in sensory evaluation of heterogeneous product types. *J. Food Sci.* 50:511-513.

Baker, R. A. 1968. Limitation of present objective techniques in sensory characterization. In *Correlation of subjective-objective methods in the study of odors and taste*. ASTM STP 440. Philadelphia: American Society for Testing and Materials.

Barnes, D. L., S. J. Harper, F. W. Bodyfelt, and M. R. McDaniel. 1991. Prediction of consumer acceptability of yogurt by sensory and analytical measures of sweetness and sourness. *J. Dairy Sci.* (submitted)

Bauman, H. E., and C. Taubert. 1984. Why quality assurance is necessary and important to plant management. *Food Technol.* 38:101-102.

Bodyfelt, F. W., J. Tobias, and G. M. Trout. 1988. *The Sensory Evaluation of Dairy Products*. New York: AVI. Van Nostrand Reinhold.

Bourne, M. C. 1977. Limitations of rheology in food texture measurements. *J. Texture Stud.* 8:219-227.

Bruhn, W. 1979. Sensory testing and analysis in quality control. Dragaco Report 9/1979:207-218.

Caplen, R. 1969. *A Practical Approach to QC*. Brandon/Systems Press.

Carlton, D. K. 1985. Plant sensory evaluation within a multi-plant international organization. *Food Technol.* 39(11):130.

Carr, B. T. 1989. An integrated system for consumer-guided product optimization. In *Product testing with consumers for research guidance*. ASTM STP 1035, ed. L. S. Wu. Philadelphia: ASTM.

Chambers, E., IV. 1990. Sensory analysis-dynamic research for today's products. *Food Technol.* 44(1):92-94.

Cochran, W. G., and G. M. Cox. 1957. *Experimental Designs*. New York: John Wiley and Sons, Inc.

Cullen, J., and J. Hollingum. 1987. *Implementing Total Quality*. UK: IFS (Publications) Ltd.

D'Agostino, R. B., A. Belanger, and R. B. D'Agostino, Jr. 1990. A Suggestion for using powerful and informative tests of normality. *The American Statistician* 44(4):316-21.

Dawson, E. H., and B. L. Harris. 1951. Sensory methods for measuring differences in food quality. *U.S. Dept. Agr., Agr. Infor. Bull.* 34:1-134.

Dove, W. F. 1947. Food acceptability-its determination and evaluation. *Food Technol.* 1:39-50.

Dziezak, J. D. 1987. Quality assurance through commercial laboratories and consultants. *Food Technol.* 41(12):110-127.

Feigenbaum, A. V. 1951. *Quality Control: Principles, Practice and Administration*. New York: McGraw-Hill Book Company, Inc.

Finney, E. E. 1969. Objective measurements for texture in foods. *J. Texture Stud.* 1:19-37.

Foster, D. 1968. Limitations of subjective measurement of odors. In *Correlation of Subjective-Objective Methods in the Study of Odors and Taste*. ASTM STP 440. Philadelphia: American Society for Testing and Materials.

Friedman, H. H., J. E. Whitney, and A. S. Szczesniak. 1963. The texturometer-a new instrument for objective measurements. *J. Food Sci.* 28:390-396.

Gacula, M. C. 1978. Analysis of incomplete block designs with reference samples in every block. *J. Food Sci.* 43:1461-1466.

Gallup Organization. 1985. *Customer perceptions concerning the quality of American products and services*. ASQC/Gallup Survey. Princeton, NJ: The Gallup Organization, Inc.

Garvin, D. A. 1984. What does product quality really mean? *Sloan Management Review*. Fall.

Garvin, D. A. 1987. Competing on the eight dimensions of quality. *Harvard Business Review* 65(6):101.

Gendron, G., and B. Burlingham. 1989. The entrepreneur of the decade. An interview with Steve Jobs. *Inc.* (April):114-128.

Gerigk, K., G. Hildebrandt, H. Stephen, and J. Wegener. 1986. Lebensmittelrechtliche Beurteilung von Leberwurst. Fleischwirtsch. 66(3):310-314.

Halberstam, D. 1984. Yes we can! *Parade Magazine* (7):4-7.

Hayes, R. 1981. Why Japanese factories work. *Harvard Business Review* (7-8):57.

Herschdoerfer, S. M. (ed). 1986. *Quality Control in the Food Industry.* New York: Academic Press.

Hinreiner, E. H. 1956. Organoleptic evaluation by industry panels-the cutting bee. *Food Technol.* 31(11):62-67.

Hubbard, M. R. 1990. *Statistical Quality Control for the Food Industry.* New York: Van Nostrand Reinhold.

Hunter, R. S. 1987. Objective methods for food appearance assessment. In *Objective Methods in Food Quality Assessment*, ed. J. G. Kapsalis. Boca Raton, FL: CRC Press.

Huntoon, R. B., and L. G. Scharpf, Jr. 1987. Quality control of flavoring materials. In *Objective Methods in Food Quality Assessment*, ed. J. G. Kapsalis. Boca Raton, FL: CRC Press.

Institute of American Poultry Industries. 1962. Chemical and bacteriological methods for the examination of egg and egg products. Chicago: Mimeo. Methodology, Egg Prod. Lab., Inst. Am. Poultry Ind.

Jacobsen, T., and R. W. Gunderson. 1986. Applied Cluster Analysis. In *Statistical Procedures in Food Research*, ed. J. R. Piggott. New York: Elsevier Science Publishing Co.

Ke, P. J., B. G. Burns, and A. D. Woyewoda. 1984. Recommended procedures and guidelines for quality evaluation of Atlantic short-fin squid (*Illex illecebrosus*). *Lebersmittelwissenschaft u. Technol.* (17):276-281.

Kepper, R. E. 1985. Quality motivation. *Food Technol.* 39 (9):51-52.

Kirk, R. E. 1968. *Experimental Design: Procedures for the Behavioral Sciences.* Belmont, CA: Wadsworth Publishing Co.

Kraemer, H. C., and S. Thiemann. 1987. *How Many Subjects?* Newbury Park, CA: SAGE Publications.

Kramer, A., and B. A. Twigg. 1970. *Quality Control for the Food Industry.* Westport, CT: The Avi Publishing Co., Inc.

Langenau, E. E. 1968. Correlation of objective-subjective methods as applied to the perfumery and cosmetics industries. In *Correlation of objective-subjective methods in the study of odors and taste.* ASTM STP 440. Philadelphia: American Society for Testing and Materials.

Larmond, E. 1976. Sensory methods-choices and limitations. In *Correlating sensory objective measurements-new methods for answering old problems.* ASTM STP 594. Philadelphia: American Society for Testing and Materials.

Levitt, D. J. 1974. The use of sensory and instrumental assessment of organoleptic characteristics in multivariate statistical methods. *J. Texture Stud.* 5:183-200.

Levy, I. 1983. Quality assurance in retailing: a new professionalism. *Quality Assurance* 9(1):3-8.

Little, A. C. 1976. Physical measurements as predictors of visual appearance. *Food Technol.* 30(10):74,76,77,80,82.

Little, A. C., and G. MacKinney. 1969. The sample as a problem. *Food Technol.* 23(1):25-28.

List, G. R. 1984. The role of the Northern Regional Research Center in the development of quality control procedures for fats and oils. *JAOCS* 61(6):1017-1022.

MacKinney, G., A. C. Little, and L. Brinner. 1966. Visual appearance of foods. *Food Technol.* 20(10):1300-1308.

MacNiece, E. H. 1958. Integrating consumer requirements in engineering specifications. In *Quality Control in Action*. AMA Management Report #9. New York: American Management Association, Inc.

Mastrian, L. K. 1985. The sensory evaluation program within a small processing operation. *Food Technol.* 39(11):127.

McNutt, K. 1988. Consumer attitudes and the quality control function. *Food Technol.* 42: 97,98,108.

Meilgaard, M., G. V. Civille, and B. T. Carr. 1987. *Sensory Evaluation Techniques*. Boca Raton, FL: CRC Press, Inc.

Merolli, A. 1980. Sensory evaluation in operations. *Food Technol.* 34(11):63-64.

Morrison, D. F. 1976. *Multivariate Statistical Methods*. New York: McGraw-Hill, Inc.

Muñoz, A. M. 1986. Development and application of texture reference scales. *J. Sensory Stud.* 1:55-83.

Nakayama, M., and C. Wessman. 1979. Application of sensory evaluation to the routine maintenance of product quality. *Food Technol.* 33:38,39,44.

Nelson, L. 1984. The Shewart control chart-tests for special causes. *J. Quality Technol.* 16(4):237-239.

Noble, A. C. 1975. Instrumental analysis of the food properties of food. *Food Technol.* 29(12):56-60.

O'Mahoney, M. 1979. Psychophysical aspects of sensory analysis of dairy products. *Dairy Sci.* 62:1954.

Ough, C. S., and G. A. Baker. 1961. Small panel sensory evaluations of wines by scoring. *Hildegardia* 30:587-619.

Pangborn, R. M. 1964. Sensory evaluation of foods: a look backward and forward. *Food Technol.* 18(9):63-67.

Pangborn, R. M. 1987. Sensory science in flavor research: achievements, needs and perspectives. In *Flavor Science and Technology*, ed. M. Martens, G. A. Dalen, and H. Russwurm, Jr. Great Britain: John Wiley & Sons, Inc.

Pangborn, R. M., and W. L. Durkley. 1964. Laboratory procedures for evaluating the sensory properties of milk. *Dairy Sci.* 26(2):55-62.

Pedraja, R. R. 1988. Role of quality assurance in the food industry: new concepts. *Food Technol.* 42(12):92-93.

Penzias, A. 1989. *Ideas and Information: Managing in a High Tech World.* W. W. Norton & Company, Inc.

Peryam, D. R., and R. Shapiro. 1955. Perception, preference, judgement-clues to food quality. *Ind. Quality Control.* 11:1-6.

Plank, R. P. 1948. A rational method for grading food quality. *Food Technol.* 2:241-251.

Platt, W. 1931. Scoring food products. *Food Indus.* 3:108-111.

Plsek, P. E. 1987. Defining quality at the marketing/development interface. *Quality Progress* 20(6):28-36.

Powers, J. J., D. R. Godwin, and R. E. Bargmann. 1977. Relations between sensory and objective measurements for quality evaluations of green beans. In *Flavor Quality-Objective Measurement*, ed. R. A. Scanlon. Washington, D.C.: American Chemical Society.

Powers, J. J., and V. N. M. Rao. 1985. Computerization of the quality assurance program. *Food Technol.* 39(11):136.

Reece, R. N. 1979. A quality assurance perspective of sensory evaluation. *Food Technol.* 33(9):37.

Rothwell, J. 1986. The quality control of frozen desserts. In *Quality Control in the Food Industry*, ed. S. M. Herschdoerfer. Vol 3, Second edition. New York: Academic Press.

Ruehrmund, M. E. 1985. Coconut as an ingredient in baking foods. *Food Technology in New Zealand.* (11):21-25.

Rutenbeck, S. K. 1985. Initiating an in-plant quality control/sensory evaluation program. *Food Technol.* 39(11):124.

SAS Institute Inc. *SAS/STAT Alseis Guide, Version 6, Fourth Edition, Volume 2.* Cary, NC: SAS Institute Inc.

Schrock, E. M., and H. L. Lefevre. 1988. *The Good and the Bad News about Quality.* New York: Marcel Dekker, Inc.

Sidel, J. L., H. Stone, and J. Bloomquist. 1981. Use and misuse of sensory evaluation in research and quality control. *Dairy Sci.* 64:2296-2302.

Sidel, J. L., H. Stone, and J. Bloomquist. 1983. Industrial approaches to defining quality. In *Sensory Quality in Foods and Beverages: Definitions, Measurement and Control*, ed. A. A. Williams and R. K. Atkin, pp. 48-57. Chichester, UK: Ellis Horwood Ltd.

Sinha, M. N., and W. Willborn. 1985. *The Management of Quality Assurance.* John Wiley and Sons.

Sjostrom, L. B. 1968. Correlation of objective-subjective methods as applied in the food field. In *Correlation of objective-subjective methods in the study of odors and taste.* ASTM STP 440. Philadelphia: American Society for Testing Materials.

Smith, G. L. 1988. Statistical analysis of sensory data. In *Sensory Analysis of Foods, Second Edition*, ed. J. R. Piggott. New York: Elsevier Science Publishing Co.

Spencer, H. W. 1977. Proceedings of Joint IFST and Food Group Society of Chemical Industry. Symposium on Sensory Quality Control.

Steber, H. 1985. Qualitatskontrolle bei Fruchtzubereitungen. Deutsche Molkerei-Zeitung 46:1547-1550.

Stevenson, S. G., M. Vaisey-Genser, and N. A. M. Eskin. 1984. Quality control in the use of deep frying oils. *JAOCS* 61(6):1102-1108.

Stone, H., and J. Sidel. 1985. *Sensory Evaluation Practices.* New York: Academic Press.

Stouffer, J. C. 1985. Coordinating sensory evaluation in a multi plant operation. *Food Technol.* 39(11):134.

Szczesniak, A. S. 1973. Instrumental methods of texture measurements. *Texture Measurements of Foods*, ed. A. Kramer and A. S. Szczesniak. Dordrecht, Holland: D. Reidel Pub. Co.

Szczesniak, A. S. 1987. Correlating sensory with instrumental texture measurements. An overview of recent developments. *J. Texture Stud.* 18:1-15.

Tompkins, M. D., and G. B. Pratt. 1959. Comparison of flavor evaluation methods for frozen citrus concentrates. *Food Technol.* 13:149-152.

Trant, A. S., R. M. Pangborn, and A. C. Little. 1981. Potential fallacy of correlating hedonic responses with physical and chemical measurements. *J. Food Sci.* 46(2):583-588.

Waltking, A. E. 1982. Progress report of the AOCS flavor nomenclature and standards committee. *JAOCS.* 59(2):116A-120A.

Wetherill, G. B. 1977. *Sampling Inspection and Quality Control, Second Edition*. New York: Chapman and Hall.

Williams, A. A. 1978. Interpretation of the sensory significance of chemical data in flavor research. *IFFA* (3/4):80–85.

Wolfe, K. A. 1979. Use of reference standards for sensory evaluation of product quality. *Food Technol*. 33:43–44.

Woollen, A. 1988. How Coca-Cola assures quality with QUACS. *Soft Drinks Management International* (3):21–24.

Index

9 781489 926555